Civil PE
Sample Examination

Michael R. Lindeburg, PE

Professional Publications, Inc. • Belmont, CA

Errata and Other Updates for This Book

At Professional Publications, we do our best to bring you error-free books. But when errors do occur, we want to make sure that you know about them so they cause as little confusion as possible.

A current list of known errata and other updates for this book is available on the PPI website at **www.ppi2pass.com**. From the website home page, click on "Errata." We update the errata page as often as necessary, so check in regularly. You will also find instructions for submitting suspected errata. We are grateful to every reader who takes the time to help us improve the quality of our books by pointing out an error.

CIVIL PE SAMPLE EXAMINATION

Current printing of this edition: 1

Printing History

edition number	printing number	update
1	1	New book.

Printed in the United States of America

Professional Publications, Inc.
1250 Fifth Avenue, Belmont, CA 94002
(650) 593-9119
www.ppi2pass.com

Library of Congress Cataloging-in-Publication Data
Lindeburg, Michael R.
    Civil PE sample examination / Michael R. Lindeburg.
        p. cm.
    Includes bibliographical reference and index.
    ISBN 1-59126-005-1
    1. Civil engineering--Examinations, questions, etc. 2. Civil engineering--Problems, exercises, etc. I. Title.

TA159.L55 2005
624′.076--dc22

                                                            2004060064

Table of Contents

PREFACE AND ACKNOWLEDGMENTS   v

CODES, HANDBOOKS, AND REFERENCES   ix

INTRODUCTION   xi

MORNING SESSION   1

AFTERNOON SESSION
   Environmental   17
   Geotechnical   25
   Structural   33
   Transportation   43
   Water Resources   53

ANSWER KEY   63

SOLUTIONS
   Morning Session   67
   Afternoon Session—Environmental   81
   Afternoon Session—Geotechnical   99
   Afternoon Session—Structural   113
   Afternoon Session—Transportation   133
   Afternoon Session—Water Resources   145

Preface and Acknowledgments

This *Civil Engineering Sample Examination* reflects the change to a "breadth and depth" (B&D) format by the National Council of Examiners for Engineering and Surveying (NCEES). NCEES developed this new B&D format in order to reduce variations in the passing percentage from exam administration to exam administration, as well as to speed up, simplify, and economize the grading and reporting processes. Another benefit, to both NCEES and the state boards, is the reduction or elimination of costly, time-consuming appeals.

The B&D format ushered in several changes to the PE exam that have been painful for examinees. The first change was that all problems on the exam became multiple choice. Long gone are the free-response essays in which you could get partial credit based on a correct method. This is a blessing in some problems, since all of the simplifying assumptions and required data must be provided. However, recent administrations of the B&D exam have demonstrated that some multiple-choice problems are more difficult and time consuming than the corresponding free-response problems.

The second change incorporated into the new format is the "no-choice" aspect. The morning breadth session consists of a wide-ranging collection of 40 problems, with no selection possible. To receive full credit for the morning, some expertise in all of the subject categories (structural and nonstructural) is required. This makes the exam more difficult than when examinees could pick 4 problems out of 10, as the abandoned format allowed.

Finally, the B&D format allows you to select (without regard to your actual area of expertise) one of several specialty categories in the afternoon depth session: structures, transportation, water resources, environmental, and geotechnical. Although there is some overlap of problems among these categories, you will be able to work mostly in an area of expertise. The benefit of the privilege, however, is partially offset by the fact that the afternoon session is also no choice, with no selection of problems possible after you have chosen your specialty.

We do a drill in competitive soccer training that is basically an endless series of sprints and rests. The players call this drill, "choke and puke." Tongue-in-cheek, I have given this sample exam the same name. You are now reading the choke-and-puke edition of the *Civil Engineering Sample Exam.*

This is not the first sample exam I have written for civil engineers, but this book is so significantly updated from my earlier sample exams that it has received a new title, and it is starting out as edition 1. This publication mimics the actual civil engineering PE licensing exam as closely in format, depth, and variety of questions as any exam ever published by Professional Publications. Therein is the difficulty. This is a very difficult sample exam. It will make you want to choke and puke. So before you get discouraged from eight hours of working this sample exam, I wanted to clarify some items:

- Although you only get six minutes per problem on the real exam, you'll need more than six minutes for many of those problems. The same is true with this sample exam. Some of the problems in this sample exam are, to be honest, 15- to 20-minute problems.

- However, there are also a number of 30-second, nonquantitative questions. They are difficult, but if you know the subject material, they'll only take you the time needed to read through them. If you don't know the subject, well,...choke and puke... you'll be furiously leafing through some of the references you have brought with you.

- The breadth of actual exam topics is intimidating. The knowledge you'll need to complete this sample exam will take you far afield from what is in any single book in your library. Even the fabled *Civil Engineering Reference Manual* isn't going to make up for a lack of experience in the afternoon portion of the exam. Nothing can.

- Luckily, you don't need to score 100%, 90%, or probably even 80% on the exam to pass. Take solace in knowing that you can get a bunch of the questions wrong. Nobody outside of exam administration is supposed to know what the required raw passing score is (it's probably lower than you would think), but it's just good to know that the score is attainable by normal geeky people—you don't need to be an idiot savant.

Engineering economics is a subject that continues to appear, here and there, once in a while. Although the subject has been eliminated from almost all of NCEES' PE exam outlines, NCEES allows that "some problems will contain aspects of engineering economics." So, although the days are gone when an examinee had to exhibit the knowledge of a tax accountant, it is still open season for such problems.

In writing this edition, I made several style and formatting decisions with which you might not agree. I used the same variable symbols as in the *Civil Engineering Reference Manual*, which was modeled after the original source documents (codes and standards).

Another decision involves the number of significant digits used in the solutions. The rules for significant digits are well known: answers cannot be any more precise than the most imprecise parameter. Field practice often uses even fewer significant digits than would be justified by the given data, in recognition of the many unknowns and assumptions incorporated into the solution. But exam review is not field work. For this publication, I assumed you would be using a calculator and would want to compare this book's solutions with the digits appearing in your calculator. I also assumed that you would retain intermediate answers in your calculator's memory, rather than enter new numbers that you already rounded. Because of these assumptions, most of the results are printed with more significant digits than can be justified. This is intentional: you'll be able to see if you're doing it the right way.

Another decision had to do with the units used in the solutions. I used common field units wherever possible, but not everywhere. As much as possible, I have tried to be consistent within this publication. Therefore, I have not used "kips" in one place and "k" in another to mean the same thing: "1000 pounds." Still, one question might be solved using "ksi," while another is solved with "psi." It all depends on the context.

In keeping with a personal preference, I was strict in differentiating between weight and mass. Thus, the units "lbf" and "lbm" are differentiated throughout, although "lb" is good enough for most civil engineering work. Similarly, I differentiated mass density (ρ) from specific weight (γ), even though the good-old "pcf" is comfortably indistinct in its meaning.

When writing and solving the numerous metric questions in this book, I used only standard SI units. For example, you won't find any "kilogram-force" units in this book, even though they might still be in use in other countries. The SI system uses the kilogram as a unit of mass, never as weight or force.

All of the problems in this sample examination are independent. NCEES makes an attempt to "decouple" problems from previous results. For example, if question 1 asks, "What is the coefficient of active earth pressure?", then question 2 will often be phrased, "Assuming the coefficient of active earth pressure is 0.30, what is the active force on the wall?" In this manner, you still have a chance on question 2 even if you get question 1 wrong. That's good.

And bad. In some problem sets, this decoupling results in you having to repeat previous steps with the new set of given data for each subsequent problem. Although some examinees have complained about the additional workload and not being able to use the interim results in their calculator stack, many others are quite comfortable with this.

Finally, just as the *Civil Engineering Reference Manual* exceeds the scope of the actual licensing exam, this sample exam does so as well. It's about 10% more difficult than the actual exam. It's still representative and realistic, just a tiny bit more difficult.

This is a sample exam. You should use it for practice, not for prediction. You should not use this examination to predict the topics on the actual exam, or (heaven forbid) to predict your own performance on the actual exam. Every exam is different. Everyone is different. Everyone has different strengths and weaknesses. Everyone has a different knowledge base and a different working speed. The passing rate for the PE exam has never been consistent. So, with each examinee being different and with each PE exam being different, it is unlikely that I would ever be able to perfectly match the complexity and level of difficulty you experience, anyway.

I don't think one engineer in a hundred is going to be concerned with my editorial decisions. However, I made them, and I wanted you to know about them so that you don't complain when you see the real exam is different.

Some of the questions in this publication are heritage questions dating back to earlier editions. However, most of the questions are brand new. None of the problems in this publication is an actual exam problem. The new problems in this edition have come out of my head and the heads of colleagues, based on the examination specifications published by NCEES.

I owe a huge debt to a number of people who kept this publication on track with the current exam and modern practice. James R. Sheetz, PE, DEE was responsible for the content of all environmental and water resources problems. Edmund Medley, PhD, PE, CEG co-authored the majority of the geotechnical material along with Pablo F. Sanz, PE, taking over from Robert H. Kim, MSCE, PE, who got the geotech ball rolling. C. Dale Buckner, PhD, PE continues to be Professional Publications' "go-to" person for dependable structural content. The amount of new structural problems he provided is substantial. Maher M. Murad, PhD authored the transportation problems.

All of our contributions were examined by technical reviewers before Professional Publications drew a single line or typeset a single word. David G. Smith, PE, PLS reviewed the transportation material; Thomas W. Schreffler, QEP, PE technically reviewed the environmental and water resources material; Thomas H. Miller, PhD, PE covered us for structural; and Edmund Medley, PhD, PE, CEG and Pablo F. Sanz, PE technically reviewed some of the early geotechnical material before becoming coauthors.

The staff that converted the technically reviewed, but raw, manuscript into a publishable work is the best trained and most proficient that Professional Publications has ever had. Managed by Sarah Hubbard, Editorial Director, and Cathy Schrott, Production Director, the following top-drawer individuals have my thanks for bringing this book to life: Sean Sullivan, Project Editor; Amy Schwertman, Illustrator; Miriam Hanes, Typesetter, and Kate Hayes, Typesetter.

As in all of my publications, I invite your comments. If you disagree with a solution, or if you think there is a better way to do something, please let me know (at www.ppi2pass.com/erratasubmit) so that I can share your comments with everyone else.

Michael R. Lindeburg, PE

Codes, Handbooks, and References

PPI lists on its website the dates of the codes, standards, regulations, and references on which NCEES has based the current exams. This *Civil Engineering Sample Exam* is also based on this information, but it is a moving target. As with engineering practice itself, where adoptions by state and local agencies often lag issuance by several years, the NCEES exams are not necessarily based on the most current codes. The exam can be years behind the most recently adopted codes, and many years behind the latest codes being published and distributed. In some cases, it may even be impossible to find copies of the documents on which the exam is based if they have gone out of print. When this book was written, the following codes and references were used.

Codes

AASHTO: *Design of Pavement Structures*, 1993, American Association of State Highway and Transportation Officials, Washington, DC

AASHTO: *Roadside Design Guide*, 2002, American Association of State Highway and Transportation Officials, Washington, DC

AASHTO: *Standard Specifications for Highway Bridges*, Seventeenth ed., 2002, American Association of State Highway and Transportation Officials, Washington, DC

AASHTO Green Book: *Policy on Geometric Design of Highways and Streets*, Fourth ed., English units version, 2001, American Association of State Highway and Transportation Officials, Washington, DC

ACI 318: *Building Code Requirements for Structural Concrete*, 2002, American Concrete Institute, Farmington Hills, MI

ACI 530/ASCE 5/TMS 402: *Building Code Requirements for Masonry Structures*, Ninth ed., 2002, published jointly by American Concrete Institute, American Society of Civil Engineers, and The Masonry Society

ACI 530.1/ASCE 6/TMS 602: *Specifications for Masonry Structures*, 2002, published jointly by American Concrete Institute, American Society of Civil Engineers, and The Masonry Society

AI: *The Asphalt Handbook, Manual MS-4*, 1989, Asphalt Institute, College Park, MD

AISC/ASD: *Manual of Steel Construction, Allowable Stress Design*, Ninth ed. (without supplements), 1989, American Institute of Steel Construction, Inc., Chicago, IL

AISC/LRFD: *Manual of Steel Construction, Load and Resistance Factor Design*, Third ed., 2001, American Institute of Steel Construction, Inc., Chicago, IL

ASCE: *Minimum Design Loads for Buildings and Other Structures*, 2002, American Society of Civil Engineers, New York, NY

HCM 2000: *Highway Capacity Manual 2000*, 2000, Transportation Research Board/National Research Council, Washington, DC

IBC: *International Building Code*, (without supplements), 2003, International Code Council, Falls Church, VA

ITE: *Traffic Engineering Handbook*, Fifth ed., 1999, Institute of Transportation Engineers, Washington, DC

MUTCD: *Manual of Uniform Traffic Control Devices for Streets and Highways*, 2001, U.S. Department of Transportation, Federal Highway Administration, Washington, DC

NDS: *National Design Specification for Wood Construction*, (ASD edition with ASD supplement), 2001, American Forest and Paper Association, Washington, DC

PCA: *Design and Control of Concrete Mixtures*, Fourteenth ed., 2002, Portland Cement Association, Skokie, IL

PCI: *PCI Design Handbook*, Fifth ed., 1999, Precast/Prestressed Concrete Institute, Chicago, IL

Introduction

ABOUT THE PE EXAM

The *Civil PE Sample Examination* provides the opportunity to practice taking an eight-hour test similar in content and format to the Principles and Practice of Engineering (PE) examination in civil engineering. The civil PE examination is an eight-hour exam divided into a morning session and an afternoon session. The morning session is known as the "breadth" exam, and the afternoon session is known as the "depth" exam. This book contains a sample breadth module and five sample depth modules—one for each subdiscipline the NCEES tests.

In the four-hour morning session, the examinee is asked to solve 40 problems from five major civil engineering subdisciplines: environmental (approximately 20% of the exam problems); geotechnical (20%); structural (20%); transportation (20%); and water resources (20%). Morning session problems are general in nature and wide-ranging in scope.

The four-hour afternoon session allows the examinee to select a depth exam module from one of five subdisciplines (environmental, geotechnical, structural, transportation, and water resources). Each depth module is made up of 40 problems. Afternoon session problems require more specialized knowledge than those in the morning session.

All problems, from both the morning and afternoon sessions, are multiple choice. They include a problem statement with all required defining information, followed by four logical choices. Only one of the four options is correct. The problems are completely independent of each other, so an incorrect choice on one problem will not carry over to subsequent problems.

This book is written in the multiple-choice exam format instituted by the NCEES. It covers all the same topic areas that appear on the exam, as provided by the NCEES.

Topics and the approximate distribution of problems on the morning session of the civil PE exam are as follows.

Environmental (20%):

- Wastewater treatment
- Biology

- Solid/hazardous waste
- Groundwater and well fields

Geotechnical (20%):

- Subsurface exploration and sampling
- Engineering properties of soils
- Soil mechanics analysis
- Shallow foundations
- Earth retaining structures

Structural (20%):

- Loadings
- Analysis
- Mechanics of materials
- Materials
- Member design

Transportation (20%):

- Traffic analysis
- Construction
- Geometric design

Water Resources (20%):

- Hydraulics
- Hydrology
- Water treatment

Topics and the approximate distribution of problems in the five depth modules of the civil PE exam afternoon session are as follows.

Environmental Module

- Environmental (65%)
- Geotechnical (10%)
- Water resources (25%)

Geotechnical Module

- Geotechnical (65%)
- Environmental (10%)

- Structural (20%)
- Transportation (5%)

Structural Module

- Structural (65%)
- Geotechnical (25%)
- Transportation (10%)

Transportation Module

- Transportation (65%)
- Geotechnical (15%)
- Water resources (20%)

Water Resources Module

- Water resources (65%)
- Environmental (25%)
- Geotechnical (10%)

According to the NCEES, exam questions related to codes and standards will be based on either (1) an interpretation of a code or standard that is presented in the exam booklet or (2) a code or standard that a committee of licensed engineers feels minimally competent engineers should know. Code information required to solve questions will be consistent with the last edition of the code issued before the year of the exam.

For further information and tips on how to prepare for the civil PE exam, consult the *Civil Engineering Reference Manual* or Professional Publications' website, www.ppi2pass.com.

HOW TO USE THIS BOOK

It is recommended that you treat this book as an exam. Do not read the questions ahead of time, and do not look at the answers until you've finished. As you work the problems, you may use the *Civil Engineering Reference Manual*. Adequate preparation, not an extensive library, is the key to success. Check with your state's board of engineering registration for any restrictions. (The PPI website, www.ppi2pass.com, has a listing of state boards.)

Prepare for the exam, read the sample exam instructions (which simulate the ones you'll receive from your exam proctor), set a timer for four hours, and take the breadth module. After a one-hour break, turn to the depth module you will select during the actual exam, set the timer, and complete the simulated afternoon session. Then, check your answers.

After taking the sample exam, review your areas of weakness and then take the exam again, but since none of the problems in the book are repeated, substitute a different depth module. Check your answers, and repeat the process for each of the depth areas. Evaluate your strengths and weaknesses, and select additional texts to supplement your weak areas (e.g., *Practice Problems for the Civil Engineering PE Exam*). Check the PPI website for the latest in exam preparation materials at www.ppi2pass.com.

The problems in this book were written to emphasize the breadth of the civil engineering field. Some may seem easy and some hard. If you are unable to answer a given question, you should review that topic area.

This book assumes that the breadth module of the PE exam will be more academic and traditional in nature, and that the depth modules will require practical, non-numerical knowledge, of the type that comes from experience.

The problems are generally similar to each other in difficulty, yet a few somewhat easier problems have been included to expose you to less-frequently examined topics.

The keys to success on the exam are to know the basics and to practice solving as many problems as possible. This book will assist you with both objectives.

Morning Session
Instructions

In accordance with the rules established by your state, you may use textbooks, handbooks, bound reference materials, and any approved battery- or solar-powered, silent calculator to work this examination. However, no blank papers, writing tablets, unbound scratch paper, or loose notes are permitted. Sufficient room for scratch work is provided in the Examination Booklet.

You are not permitted to share or exchange materials with other examinees. However, the books and other resources used in this morning session may be changed prior to the afternoon session.

You will have four hours in which to work this session of the examination. Your score will be determined by the number of questions that you answer correctly. There is a total of 40 questions. All 40 questions must be worked correctly in order to receive full credit on the exam. There are no optional questions. Each question is worth one point. The maximum possible score for this section of the examination is 40 points.

Partial credit is not available. No credit will be given for methodology, assumptions, or work written in your Examination Booklet.

Record all of your answers on the Answer Sheet. No credit will be given for answers marked in the Examination Booklet. Mark your answers with the pencil provided to you. Marks must be dark and must completely fill the bubbles. Record only one answer per question. If you mark more than one answer, you will not receive credit for the question. If you change an answer, be sure the old bubble is erased completely; incomplete erasures may be misinterpreted as answers.

If you finish early, check your work and make sure that you have followed all instructions. After checking your answers, you may turn in your Examination Booklet and Answer Sheet and leave the examination room. Once you leave, you will not be permitted to return to work or change your answers.

When permission has been given by your proctor, break the seal on the Examination Booklet. Check that all pages are present and legible. If any part of your Examination Booklet is missing, your proctor will issue you a new Booklet.

Do not work any questions from the Afternoon Session during the first four hours of this exam.

WAIT FOR PERMISSION TO BEGIN

Name: _____
　　　　Last　　　　　　First　　　　Middle Initial

Examinee number: _____

Examination Booklet number:_____

Principles and Practice of Engineering Examination

Morning Session
Sample Examination

1. Ⓐ Ⓑ Ⓒ Ⓓ 11. Ⓐ Ⓑ Ⓒ Ⓓ 21. Ⓐ Ⓑ Ⓒ Ⓓ 31. Ⓐ Ⓑ Ⓒ Ⓓ
2. Ⓐ Ⓑ Ⓒ Ⓓ 12. Ⓐ Ⓑ Ⓒ Ⓓ 22. Ⓐ Ⓑ Ⓒ Ⓓ 32. Ⓐ Ⓑ Ⓒ Ⓓ
3. Ⓐ Ⓑ Ⓒ Ⓓ 13. Ⓐ Ⓑ Ⓒ Ⓓ 23. Ⓐ Ⓑ Ⓒ Ⓓ 33. Ⓐ Ⓑ Ⓒ Ⓓ
4. Ⓐ Ⓑ Ⓒ Ⓓ 14. Ⓐ Ⓑ Ⓒ Ⓓ 24. Ⓐ Ⓑ Ⓒ Ⓓ 34. Ⓐ Ⓑ Ⓒ Ⓓ
5. Ⓐ Ⓑ Ⓒ Ⓓ 15. Ⓐ Ⓑ Ⓒ Ⓓ 25. Ⓐ Ⓑ Ⓒ Ⓓ 35. Ⓐ Ⓑ Ⓒ Ⓓ
6. Ⓐ Ⓑ Ⓒ Ⓓ 16. Ⓐ Ⓑ Ⓒ Ⓓ 26. Ⓐ Ⓑ Ⓒ Ⓓ 36. Ⓐ Ⓑ Ⓒ Ⓓ
7. Ⓐ Ⓑ Ⓒ Ⓓ 17. Ⓐ Ⓑ Ⓒ Ⓓ 27. Ⓐ Ⓑ Ⓒ Ⓓ 37. Ⓐ Ⓑ Ⓒ Ⓓ
8. Ⓐ Ⓑ Ⓒ Ⓓ 18. Ⓐ Ⓑ Ⓒ Ⓓ 28. Ⓐ Ⓑ Ⓒ Ⓓ 38. Ⓐ Ⓑ Ⓒ Ⓓ
9. Ⓐ Ⓑ Ⓒ Ⓓ 19. Ⓐ Ⓑ Ⓒ Ⓓ 29. Ⓐ Ⓑ Ⓒ Ⓓ 39. Ⓐ Ⓑ Ⓒ Ⓓ
10. Ⓐ Ⓑ Ⓒ Ⓓ 20. Ⓐ Ⓑ Ⓒ Ⓓ 30. Ⓐ Ⓑ Ⓒ Ⓓ 40. Ⓐ Ⓑ Ⓒ Ⓓ

Morning Session

1. A complete-mix activated sludge process is used to treat 8 MGD of brewery wastes with a COD of 1800 mg/L. The nonbiodegradable fraction is 110 mg/L COD. The biochemical reaction is pseudo-first order. The substrate utilization rate constant based on mixed liquor volatile suspended solids (MLVSS) is 0.6 L/g·h at 20°C. The design mixed liquor suspended solids (MLSS) is 2500 mg/L, and the solids are 75% volatile. Activated sludge is returned directly to the reactor for a design effluent COD of 200 mg/L. The reactor volume is most nearly

(A) 200,000 ft^3
(B) 400,000 ft^3
(C) 500,000 ft^3
(D) 700,000 ft^3

2. The design flow will be 15 MGD for a wastewater treatment plant to process influent with a raw suspended solids content of 500 mg/L. The target suspended solids concentration is 150 mg/L as the flow leaves the primary clarifier treatment. A pilot plant has been built with the results shown. Two cylindrical clarifier units are required for reliability. Which of the following configurations will satisfy the design criteria for each unit?

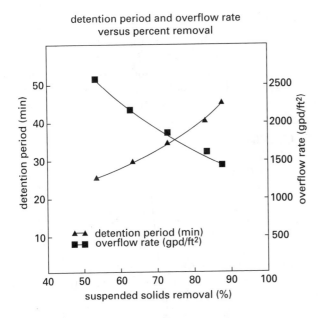

detention period and overflow rate
versus percent removal

(A) diameter of 60 ft, depth of 12 ft
(B) diameter of 70 ft, depth of 6 ft
(C) diameter of 85 ft, depth of 8 ft
(D) diameter of 100 ft, depth of 7 ft

3. Which of the following are communicable disease agents that can be transmitted through water supply and wastewater?

I. *Salmonella typhi*

II. *Escherichia coli*

III. *Giardia lamblia*

IV. *Sphaerotilis natans*

V. *Poliovirus*

VI. *Ceriodaphnia dubia*

VII. *Endamoeba histolytica*

VIII. *Pimephales promelas*

(A) I, II, V, VII
(B) I, III, V, VII
(C) III, IV, VI, VIII
(D) V, VI, VII, VIII

4. An industrial wastewater is tested for chronic toxicity, yielding the results in the table for the following freshwater species.

species	exposure (days)	control survival (%)	NOEC (% effluent)	LOEC (% effluent)
Pimephales promelas	7	99	5.0	8.0
Ceriodaphnia dubia	7	98	2.0	4.5
Rainbow trout	7	100	12.0	18.0

If the dilution in the mixing zone where the waste is discharged is 100:1, the criterion continuous concentration (CCC) for the test is most nearly

(A) 0.08 chronic toxicity unit
(B) 0.20 chronic toxicity unit
(C) 0.50 chronic toxicity unit
(D) 1.2 chronic toxicity units

5. A freshwater stream exhibits the characteristics shown in the following table. The results come from three zones studied during a 30 d period.

| | condition | | |
parameter	zone 1	zone 2	zone 3
dissolved oxygen	4.5 mg/L	90% saturation	0.5 mg/L
water temperature	13°C	10°C	23°C
total coliform			
(geometric mean)	400/100 mL	100/100 mL	5000/100 mL
biochemical oxygen			
demand	3 mg/L	1 mg/L	6 mg/L
Spaerotilus natans	slight	absent	abundant
midge larvae	low density	abundant	low density
sulfide odors	absent	absent	present

Which of the following statements are true relative to each zone and the ecology of clean and polluted water?

I. Zone 1 would be characterized as an oligosaprobic zone and would be representative of highly polluted water.

II. Zone 2 represents relatively clean water and would be classified as an oligosaprobic zone.

III. Zone 3 represents a highly polluted water and would be classified as a polysaprobic zone.

IV. Zones 1 and 3 are representative of highly polluted water and would be classified as mesosaprobic zones.

(A) I, II
(B) II, III, IV
(C) II, III
(D) III, IV

6. A solid waste facility recovers salable metals, glass, and other materials from municipal solid waste. The remaining solid waste is incinerated. The combustion heat generates steam for turbines, which in turn drive electrical generators. The electrical generation plant consists of three power modules, each containing a furnace, a steam generator, a turbine, and an electrical generator. Each power module burns 180 U.S. tons of solid waste and 190 U.S. tons of sludge each day, 7 days a week.

When received by the electrical generation plant, the solid waste contains 5% moisture and 15% material that

is removed by an air classifier. The moisture and classified material are removed prior to combustion. If per capita solid waste generation is 7 lbm/day, the population served by the solid waste facility is most nearly

(A) 100,000 people
(B) 200,000 people
(C) 300,000 people
(D) 400,000 people

7. A community of 25,000 people generates 8 lbm of solid waste per person per day, which is disposed of in a landfill. The landfill's design specifications include an in-place density of solid waste of 1200 lbm/yd^3, a soil cover ratio of four parts solid waste to one part cover, and an in-place soil density of 130 lbm/ft^3. The total mass of soil required for 1 yr of operation is most nearly

(A) 15,000 U.S. tons
(B) 25,000 U.S. tons
(C) 35,000 U.S. tons
(D) 45,000 U.S. tons

8. A contaminant plume of tetrachloroethylene moves toward a river 2 km away. The coefficient of retardation for tetrachloroethylene is given in the table.

travel time (d)	coefficient of retardation
100	0.25
400	0.19
>640	0.17

The depth of the plume is 50 m, and the groundwater elevation is 40 m above the river level. The aquifer is predominately clay, sand, and gravel, with a Darcy coefficient of 0.1 m/d and a porosity of 0.25. What is most nearly the time it will take the contaminant to reach the river?

(A) 4000 yr
(B) 6000 yr
(C) 8000 yr
(D) 10,000 yr

9. A gravity retaining wall is holding soil backfill with the properties shown. The total active force per unit length of the wall is most nearly

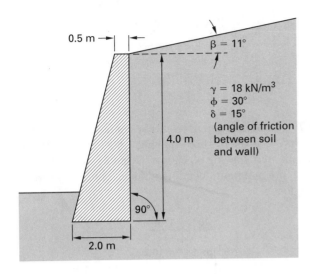

(A) 40 kN/m
(B) 50 kN/m
(C) 70 kN/m
(D) 80 kN/m

10. A sand fill is spread on top of organic silt as shown. Assume that the fill is infinite in extent. At the end of the consolidation process, the increase in vertical effective stress at point A due to the placement of the sand fill is most nearly

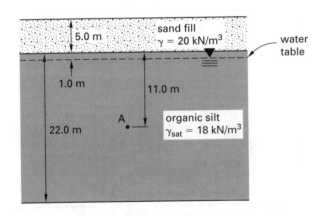

(A) 0 kN/m²
(B) 60 kN/m²
(C) 100 kN/m²
(D) 120 kN/m²

11. An impermeable dam impounds water over a soil with properties as shown. The width of the dam across the water flow is 300 m. The quantity of seepage under the dam is most nearly

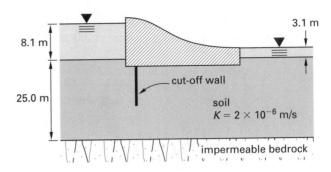

(A) 1.5×10^{-6} m³/s
(B) 2.5×10^{-6} m³/s
(C) 1.5×10^{-3} m³/s
(D) 2.5×10^{-3} m³/s

12. A square foundation supports a column load of 800 kN. The soil beneath the footing is generally homogeneous. If the foundation bearing pressure from this load is reduced from 400 kPa to 100 kPa (the column load remaining constant), the change in stress at a depth of 3 m below the foundation center will be most nearly

(A) a decrease in stress of 20 kPa
(B) a decrease in stress of 10 kPa
(C) an increase in stress of 10 kPa
(D) an increase in stress of 20 kPa

13. A retaining wall is shown. For the given conditions, the factor of safety against overturning is most nearly

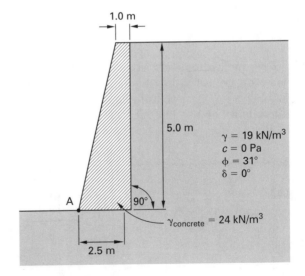

(A) 1.3
(B) 2.3
(C) 2.6
(D) 2.8

14. A slope with the soil properties given is shown. The cohesive factor of safety for the stability of this slope is most nearly

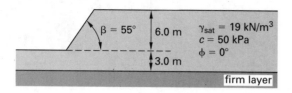

(A) 1.0
(B) 1.5
(C) 2.0
(D) 2.5

15. The results from a series of direct shear tests on a sandy soil are shown in the table.

test number	normal stress (kPa)	shear stress (kPa)
1	50	36
2	150	105
3	250	175

The principal stresses on the failure plane for test 2 are most nearly

(A) $\bar{\sigma}_1 = 100$ kPa and $\bar{\sigma}_3 = 250$ kPa
(B) $\bar{\sigma}_1 = 100$ kPa and $\bar{\sigma}_3 = 350$ kPa
(C) $\bar{\sigma}_1 = 150$ kPa and $\bar{\sigma}_3 = 390$ kPa
(D) $\bar{\sigma}_1 = 150$ kPa and $\bar{\sigma}_3 = 490$ kPa

16. Sieve and hydrometer testing shows that a soil has the following grain size distribution. The material passing through a no. 40 sieve has a liquid limit of 34 and a plasticity index of 13. The AASHTO classification for this soil is

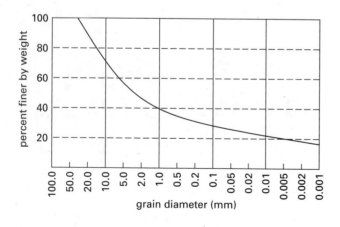

(A) A-2-6 (0)
(B) A-2-6 (1)
(C) A-2-7 (0)
(D) A-2-7 (1)

17. A simply supported girder spans 80 ft and is subjected to a set of three moving wheel loads with magnitude and spacing as shown. What is most nearly the absolute maximum bending moment caused by the moving loads?

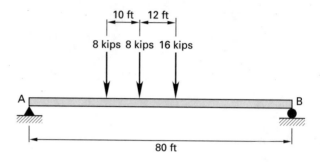

(A) 510 ft-kips
(B) 590 ft-kips
(C) 650 ft-kips
(D) 740 ft-kips

18. What is most nearly the compressive force in member CD in the truss shown, where tension is positive and compression is negative?

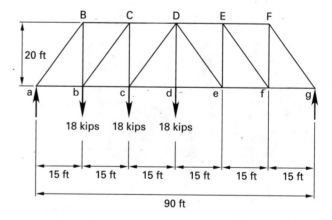

(A) −90 kips
(B) −60 kips
(C) −40 kips
(D) −20 kips

19. A solid masonry column has a cross section measuring 32 in × 32 in. The column is subjected to axial compression force of 110 kips, which includes the weight of the column, with an eccentricity of 5 in about one axis. What is most nearly the maximum axial compression stress caused by this loading?

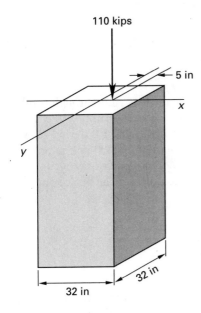

110 kips

5 in

x

y

32 in

32 in

32 in

(A) 0.2 kips/in^2
(B) 1.3 kips/in^2
(C) 2.4 kips/in^2
(D) 3.0 kips/in^2

20. A singly reinforced concrete beam has the cross section shown. The concrete is normal weight with a specified compressive strength of 4000 psi and is reinforced with four no. 10, grade 60 rebars. The design moment strength of the section is most nearly

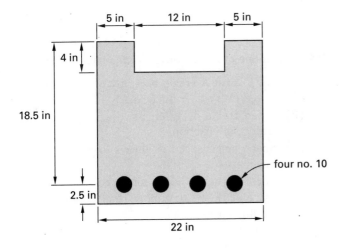

5 in   12 in   5 in

4 in

18.5 in

2.5 in

22 in

four no. 10

(A) 300 ft-kips
(B) 340 ft-kips
(C) 380 ft-kips
(D) 440 ft-kips

21. A 14 in × 14 in reinforced concrete column bears on a square spread footing that is 8.5 ft × 8.5 ft in plan, has an overall thickness of 20 in, and is reinforced with no. 8 rebars in each direction. The footing is constructed of 3000 psi normal weight concrete in accordance with the ACI 318 specification. The maximum concentric design axial force that could be supported by the footing based on its punching shear resistance is most nearly

(A) 350 kips
(B) 400 kips
(C) 450 kips
(D) 500 kips

22. An eccentrically loaded connection is made using high-strength bolts of the same size in the arrangement shown. Based on linear elastic theory, the maximum shear that occurs in the fastener group caused by the applied force is most nearly

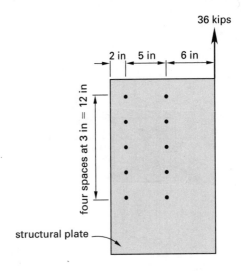

36 kips

2 in   5 in   6 in

four spaces at 3 in = 12 in

structural plate

(A) 4 kips
(B) 8 kips
(C) 10 kips
(D) 20 kips

23. A flexible plywood diaphragm spans shearwalls located on lines A, B, and C, and is subjected to a lateral wind load of 320 lbf/ft acting in the direction shown. The maximum chord force created in the diaphragm is most nearly

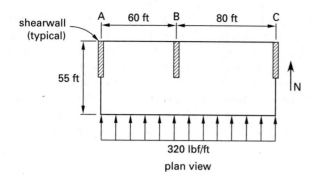

plan view

(A) 4200 lbf
(B) 4700 lbf
(C) 9600 lbf
(D) 12,000 lbf

24. Loads on a highly restrained connection result in a state of stress having equal tensile stresses on three orthogonal faces. The connection is made by welding a ductile structural steel using an appropriate electrode. Given that loads increase until failure initiates at the stressed point, the resulting failure would be best described as

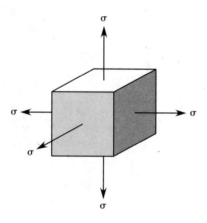

(A) ductile
(B) shear slip
(C) cleavage
(D) large deformation

25. The maximum service flow rate of a four-lane freeway is 1350 pcphpl, and the free-flow speed is 65 mph. The level of service (LOS) at which the freeway operates is most nearly

(A) LOS B
(B) LOS C
(C) LOS D
(D) LOS E

26. An unsignalized T-intersection's three legs are each two way. What is most nearly the total number of conflict points, including crossing, merging, and diverging?

(A) 3
(B) 6
(C) 9
(D) 12

27. A four-lane freeway runs through rural areas. Each lane is 11 ft wide. A recent traffic study for a particular portion of the daily commute period shows the directional weekday volume is 2400 vph in one direction. An average of 750 vehicles passes by during the busiest 15 min. What is most nearly the peak hour factor (PHF)?

(A) 0.31
(B) 0.80
(C) 0.94
(D) 1.0

28. The relationship between the average travel speed, S, in mph and the density, D, in vpm for an urban road is given by the following relationship.

$$S = 65 \ \frac{\text{mi}}{\text{hr}} - \left(0.42 \ \frac{\dfrac{\text{mi}}{\text{hr}}}{\dfrac{\text{veh}}{\text{mi}}} \right) D$$

Assume undersaturated traffic conditions. When the average travel speed is 50 mph, the flow rate is most nearly

(A) 160 vph
(B) 1800 vph
(C) 2300 vph
(D) 3200 vph

29. Which one of the following statements is true?

(A) ADT is the average of 24 hr traffic counts collected every day in the year.
(B) Fixed traffic delay on roadways is caused by traffic side friction.
(C) Space mean speed is always lower than time mean speed.
(D) Local streets provide more access than mobility, and they carry more than 80% of travel volume nationwide.

30. A car is traveling on a two-lane rural road at 45 mph. The road grade is 5% downhill. A deer appears in front of the car and starts to cross the road. What is most nearly the distance the car needs in order to stop in time to avoid hitting the deer?

(A) 150 ft
(B) 170 ft
(C) 390 ft
(D) 420 ft

(A) cut: 45 yd^3
(B) fill: 60 yd^3
(C) fill: 100 yd^3
(D) cut: 150 yd^3

31. In the following table, ADT data for traffic movements between four locations are given. Points A, B, C, and D represent the locations along various straight highway sections, as shown in the illustration. AB represents the number of daily trips from location A to location B. BA represents the number of daily trips from B to A. Other combinations of A, B, C, and D are interpreted similarly. An interchange is proposed to accommodate the traffic to and from all locations. What is the most suitable type of interchange?

turning movement	volume (vpd)
AB	18,500
AC	25
AD	15
BA	17,000
BD	30
BC	10
CD	90
CA	15
CB	20
DC	120
DB	25
DA	20

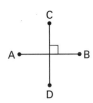

(A) cloverleaf, full
(B) cloverleaf, partial
(C) diamond
(D) directional

32. The total cut area and total fill area for two stations (1 and 2) along a roadway are as follows.

station	total cut area (ft^2)	total fill area (ft^2)
1	9	24
2	15	32

Use the average end area method for earthwork computations using 100 ft stations. What is most nearly the net earthwork volume?

33. A culvert system is being designed to pass under a major highway. The culvert system must be able to protect the highway from runoff from a 1 in storm. The following information has been derived from a storm that produced runoff over a 2 hr period.

drainage area	43 mi^2
flood hydrograph peak discharge	9300 ft^3/sec
flood hydrograph volume	3260 ac-ft

What is most nearly the 2 hr unit hydrograph peak discharge?

(A) 2300 ft^3/in-sec
(B) 3300 ft^3/in-sec
(C) 4800 ft^3/in-sec
(D) 6500 ft^3/in-sec

34. A spray system is designed as part of an industrial process. A 0.8 in diameter nozzle connected to a 3.15 in diameter pipe will provide a spray velocity of 98 ft/sec. The coefficients of velocity and contraction are 0.95 and 0.80, respectively. The water temperature is 160°F. The pressure required at the entrance to the nozzle is most nearly

(A) 46 psi
(B) 54 psi
(C) 70 psi
(D) 77 psi

35. An irrigation canal must supply 1060 ft^3/sec of water at a uniform depth of 5 ft. The canal has a 10 ft wide, rectangular cross section and is constructed of brick to make use of local building materials. The slope of the canal that will meet the supply conditions is most nearly

(A) 1.0%
(B) 1.5%
(C) 2.3%
(D) 3.4%

36. A small basin consists of the cover types given in the following table.

unit	area (ft^2)	cover type
1	200,000	open space with 80% grass, high infiltration
2	300,000	residential, $^1/_3$ ac, moderate infiltration
3	180,000	paved roads and parking

According to the NRCS method, the soil storage capacity is most nearly

(A) 1.0 in
(B) 2.5 in
(C) 3.0 in
(D) 4.5 in

37. A rapid-mixing basin is designed for a water treatment plant. The preliminary design is given in the table.

parameter	value
basin configuration	square
basin depth	1.5 times width
design flow	3.5 ft^3/sec
design velocity gradient	850 ft/sec-ft
design detention time	50 sec
temperature	70°F
impeller power number, laminar flow	65
impeller power number, turbulent flow	5.75
rotational speed	3 rev/sec

The design impeller diameter is most nearly

(A) 1.3 ft
(B) 1.6 ft
(C) 2.0 ft
(D) 2.3 ft

38. An existing water treatment plant is analyzed for deficiencies in order to improve performance for suspended solids removal with alum and lime. The analysis determined that aluminum hydroxide sludge is formed at 30 mg/L. For a flow of 0.5 m^3/s, the stoichiometric alum dose is most nearly

(A) 3000 kg/d
(B) 4000 kg/d
(C) 5000 kg/d
(D) 6000 kg/d

39. A plain sedimentation tank removes 100% of a sandy material with a mean specific gravity of 2.2, a mean diameter of 6.5×10^{-5} ft, and an operating temperature of 90°F. The system has a detention time of 2.5 hr and a flow of 18 ft^3/sec. The area and depth of the tank, respectively, are most nearly

(A) 10,000 ft^2; 13 ft
(B) 12,000 ft^2; 7 ft
(C) 14,000 ft^2; 16 ft
(D) 16,000 ft^2; 10 ft

40. Which of the following statements are true for chlorine disinfection of water for public water supply use?

I. The disinfection effectiveness is pH dependent.

II. Removal efficiencies for viruses are related to concentration, contact time, and chlorine demand.

III. Chlorine offers the additional benefit over other disinfectants of residual protection in the distribution system.

IV. A slow sand filter provides no additional benefit for chlorine disinfection.

(A) I, II, IV
(B) I, III
(C) II, III
(D) II, III, IV

STOP!

DO NOT CONTINUE!

This concludes the Morning Session of the examination. If you finish early, check your work and make sure that you have followed all instructions. After checking your answers, you may turn in your examination booklet and answer sheet and leave the examination room. Once you leave, you will not be permitted to return to work or change your answers.

Afternoon Session
Instructions

In accordance with the rules established by your state, you may use textbooks, handbooks, bound reference materials, and any approved battery- or solar-powered, silent calculator to work this examination. However, no blank papers, writing tablets, unbound scratch paper, or loose notes are permitted. Sufficient room for scratch work is provided in the Examination Booklet.

You are not permitted to share or exchange materials with other examinees. However, the books and other resources used in this afternoon session do not have to be the same as were used in the morning session.

This portion of the examination is divided into five depth modules. You may select any one of the modules, regardless of your work experience. However, you may not "jump around" and solve questions from more than one module.

You will have four hours in which to work this session of the examination. Your score will be determined by the number of questions you answer correctly. There is a total of 40 questions in each depth module. All 40 questions in the module selected must be worked correctly in order to receive full credit on the exam. There are no optional questions. Each question is worth one point. The maximum possible score for this section of the examination is 40 points.

Partial credit is not available. No credit will be given for methodology, assumptions, or work written in your Examination Booklet.

Record all of your answers on the Answer Sheet. No credit will be given for answers marked in the Examination Booklet. Mark your answers with the pencil provided to you. Marks must be dark and must completely fill the bubbles. Record only one answer per question. If you mark more than one answer, you will not receive credit for the question. If you change an answer, be sure the old bubble is erased completely; incomplete erasures may be misinterpreted as answers.

If you finish early, check your work and make sure that you have followed all instructions. After checking your answers, you may turn in your Examination Booklet and Answer Sheet and leave the examination room. Once you leave, you will not be permitted to return to work or change your answers.

When permission has been given by your proctor, break the seal on the Examination Booklet. Check that all pages are present and legible. If any part of your Examination Booklet is missing, your proctor will issue you a new Booklet.

Do not work any questions from the Morning Session during the second four hours of this exam.

WAIT FOR PERMISSION TO BEGIN

Name: _____
        Last   First   Middle Initial

Examinee number: _____

Examination Booklet number:_____

Depth Module to be graded: _____

Principles and Practice of Engineering Examination

Afternoon Session
Sample Examination

Depth Modules

Environmental 17
Geotechnical 25
Structural 33
Transportation 43
Water Resources 53

Environmental

41. (A) (B) (C) (D)   51. (A) (B) (C) (D)   61. (A) (B) (C) (D)   71. (A) (B) (C) (D)
42. (A) (B) (C) (D)   52. (A) (B) (C) (D)   62. (A) (B) (C) (D)   72. (A) (B) (C) (D)
43. (A) (B) (C) (D)   53. (A) (B) (C) (D)   63. (A) (B) (C) (D)   73. (A) (B) (C) (D)
44. (A) (B) (C) (D)   54. (A) (B) (C) (D)   64. (A) (B) (C) (D)   74. (A) (B) (C) (D)
45. (A) (B) (C) (D)   55. (A) (B) (C) (D)   65. (A) (B) (C) (D)   75. (A) (B) (C) (D)
46. (A) (B) (C) (D)   56. (A) (B) (C) (D)   66. (A) (B) (C) (D)   76. (A) (B) (C) (D)
47. (A) (B) (C) (D)   57. (A) (B) (C) (D)   67. (A) (B) (C) (D)   77. (A) (B) (C) (D)
48. (A) (B) (C) (D)   58. (A) (B) (C) (D)   68. (A) (B) (C) (D)   78. (A) (B) (C) (D)
49. (A) (B) (C) (D)   59. (A) (B) (C) (D)   69. (A) (B) (C) (D)   79. (A) (B) (C) (D)
50. (A) (B) (C) (D)   60. (A) (B) (C) (D)   70. (A) (B) (C) (D)   80. (A) (B) (C) (D)

Geotechnical

81. (A) (B) (C) (D)   91. (A) (B) (C) (D)   101. (A) (B) (C) (D)   111. (A) (B) (C) (D)
82. (A) (B) (C) (D)   92. (A) (B) (C) (D)   102. (A) (B) (C) (D)   112. (A) (B) (C) (D)
83. (A) (B) (C) (D)   93. (A) (B) (C) (D)   103. (A) (B) (C) (D)   113. (A) (B) (C) (D)
84. (A) (B) (C) (D)   94. (A) (B) (C) (D)   104. (A) (B) (C) (D)   114. (A) (B) (C) (D)
85. (A) (B) (C) (D)   95. (A) (B) (C) (D)   105. (A) (B) (C) (D)   115. (A) (B) (C) (D)
86. (A) (B) (C) (D)   96. (A) (B) (C) (D)   106. (A) (B) (C) (D)   116. (A) (B) (C) (D)
87. (A) (B) (C) (D)   97. (A) (B) (C) (D)   107. (A) (B) (C) (D)   117. (A) (B) (C) (D)
88. (A) (B) (C) (D)   98. (A) (B) (C) (D)   108. (A) (B) (C) (D)   118. (A) (B) (C) (D)
89. (A) (B) (C) (D)   99. (A) (B) (C) (D)   109. (A) (B) (C) (D)   119. (A) (B) (C) (D)
90. (A) (B) (C) (D)   100. (A) (B) (C) (D)   110. (A) (B) (C) (D)   120. (A) (B) (C) (D)

Structural

121. (A) (B) (C) (D)   131. (A) (B) (C) (D)   141. (A) (B) (C) (D)   151. (A) (B) (C) (D)
122. (A) (B) (C) (D)   132. (A) (B) (C) (D)   142. (A) (B) (C) (D)   152. (A) (B) (C) (D)
123. (A) (B) (C) (D)   133. (A) (B) (C) (D)   143. (A) (B) (C) (D)   153. (A) (B) (C) (D)
124. (A) (B) (C) (D)   134. (A) (B) (C) (D)   144. (A) (B) (C) (D)   154. (A) (B) (C) (D)
125. (A) (B) (C) (D)   135. (A) (B) (C) (D)   145. (A) (B) (C) (D)   155. (A) (B) (C) (D)
126. (A) (B) (C) (D)   136. (A) (B) (C) (D)   146. (A) (B) (C) (D)   156. (A) (B) (C) (D)
127. (A) (B) (C) (D)   137. (A) (B) (C) (D)   147. (A) (B) (C) (D)   157. (A) (B) (C) (D)
128. (A) (B) (C) (D)   138. (A) (B) (C) (D)   148. (A) (B) (C) (D)   158. (A) (B) (C) (D)
129. (A) (B) (C) (D)   139. (A) (B) (C) (D)   149. (A) (B) (C) (D)   159. (A) (B) (C) (D)
130. (A) (B) (C) (D)   140. (A) (B) (C) (D)   150. (A) (B) (C) (D)   160. (A) (B) (C) (D)

Transportation

161. (A) (B) (C) (D)
162. (A) (B) (C) (D)
163. (A) (B) (C) (D)
164. (A) (B) (C) (D)
165. (A) (B) (C) (D)
166. (A) (B) (C) (D)
167. (A) (B) (C) (D)
168. (A) (B) (C) (D)
169. (A) (B) (C) (D)
170. (A) (B) (C) (D)

171. (A) (B) (C) (D)
172. (A) (B) (C) (D)
173. (A) (B) (C) (D)
174. (A) (B) (C) (D)
175. (A) (B) (C) (D)
176. (A) (B) (C) (D)
177. (A) (B) (C) (D)
178. (A) (B) (C) (D)
189. (A) (B) (C) (D)
180. (A) (B) (C) (D)

181. (A) (B) (C) (D)
182. (A) (B) (C) (D)
183. (A) (B) (C) (D)
184. (A) (B) (C) (D)
185. (A) (B) (C) (D)
186. (A) (B) (C) (D)
187. (A) (B) (C) (D)
188. (A) (B) (C) (D)
189. (A) (B) (C) (D)
190. (A) (B) (C) (D)

191. (A) (B) (C) (D)
192. (A) (B) (C) (D)
193. (A) (B) (C) (D)
194. (A) (B) (C) (D)
195. (A) (B) (C) (D)
196. (A) (B) (C) (D)
197. (A) (B) (C) (D)
198. (A) (B) (C) (D)
199. (A) (B) (C) (D)
200. (A) (B) (C) (D)

Water Resources

201. (A) (B) (C) (D)
202. (A) (B) (C) (D)
203. (A) (B) (C) (D)
204. (A) (B) (C) (D)
205. (A) (B) (C) (D)
206. (A) (B) (C) (D)
207. (A) (B) (C) (D)
208. (A) (B) (C) (D)
209. (A) (B) (C) (D)
210. (A) (B) (C) (D)

211. (A) (B) (C) (D)
212. (A) (B) (C) (D)
213. (A) (B) (C) (D)
214. (A) (B) (C) (D)
215. (A) (B) (C) (D)
216. (A) (B) (C) (D)
217. (A) (B) (C) (D)
218. (A) (B) (C) (D)
219. (A) (B) (C) (D)
220. (A) (B) (C) (D)

221. (A) (B) (C) (D)
222. (A) (B) (C) (D)
223. (A) (B) (C) (D)
224. (A) (B) (C) (D)
225. (A) (B) (C) (D)
226. (A) (B) (C) (D)
227. (A) (B) (C) (D)
228. (A) (B) (C) (D)
229. (A) (B) (C) (D)
230. (A) (B) (C) (D)

231. (A) (B) (C) (D)
232. (A) (B) (C) (D)
233. (A) (B) (C) (D)
234. (A) (B) (C) (D)
235. (A) (B) (C) (D)
236. (A) (B) (C) (D)
237. (A) (B) (C) (D)
238. (A) (B) (C) (D)
239. (A) (B) (C) (D)
240. (A) (B) (C) (D)

Afternoon Session
Environmental

41. Wastewater design flow for a wastewater treatment plant is to be based on population for domestic sewage, plus industrial wastewater, storm water, and infiltration. The parameters are given in the following table.

parameter	value
population	65,000 people
industrial area	200 ac
peak daily flow	200% of annual average
excess infiltration rate (above normal)	600 gal/day-in-mi
collection system properties	5 mi of 24 in pipe
	16 mi of 12 in pipe
	22 mi of 8 in pipe
industrial waste	10,000 gal/ac-day

The design maximum daily flow is most nearly

(A) 10 MGD
(B) 15 MGD
(C) 20 MGD
(D) 25 MGD

42. A state regulatory agency has established the following criteria for design of primary clarifiers for municipal wastewater treatment plants.

parameter	value
overflow rate, peak hour	2000 gal/day-ft^2
overflow rate, maximum daily	800 gal/day-ft^2
sidewall depth, minimum	8 ft
weir loading rate, peak hour	35,000 gal/day-ft
detention time, minimum	60 min

The annual average design flow for the plant is 3.5 MGD. The peaking factors relative to the annual average design flow are 2.0 for the maximum day and 4.0 for the peak hour. One clarifier is to be used. The diameter of the clarifier to the next higher 10 ft increment is most nearly

(A) 100 ft
(B) 110 ft
(C) 120 ft
(D) 130 ft

43. A complete mix-activated sludge wastewater treatment plant is being studied. The characteristics of the plant are given in the following table.

parameter	value
volatile fraction of MLSS	70%
influent flow	200 L/s
MLSS	2500 mg/L
influent soluble BOD$_5$	210 mg/L
effluent soluble BOD$_5$	10 mg/L
waste rate from recycle line	20 L/s
volume of aeration	4300 m^3
SVI in recycle line	100 mL/g

The mean cell residence time is most nearly

(A) 8 h
(B) 10 h
(C) 15 h
(D) 17 h

44. A secondary clarifier is to be designed using the criteria in the following table.

parameter	value
maximum overflow rate	600 gal/day-ft^2
minimum detention time	90 min
maximum weir loading rate	10,000 gal/day-ft
design peak flow	2 MGD
minimum depth	8 ft

What is most nearly the diameter of a single basin that would meet the design criteria?

(A) 45 ft
(B) 55 ft
(C) 65 ft
(D) 75 ft

45. A wastewater treatment plant will use alum to remove 10 mg/L of phosphorus from a flow of 400 L/s. A pilot test determined that 50% above theoretical requirements for alum are needed to effectively remove the phosphorus. Reference data for the alum and phosphorus are given in the following table.

parameter	value
molecular weight of alum	666.7 g/mol
formula for liquid alum	$Al_2(SO_4)_3 \cdot 18H_2O$
alum strength	49%
concentration of alum solution	1.400 kg/L

The volume of alum solution required is most nearly

- (A) 6000 L/d
- (B) 8000 L/d
- (C) 10 000 L/d
- (D) 12 000 L/d

46. An activated-sludge wastewater treatment plant receives 400 L/s raw wastewater with 280 mg/L BOD_5 and 220 mg/L total suspended solids (TSS). The final effluent is 20 mg/L BOD_5 and 20 mg/L TSS. The primary clarifier removes 30% BOD_5 and 75% TSS. The cell yield in the aeration tanks is 60 kg suspended solids produced per 100 kg of BOD_5 removed. No BOD is removed through the secondary clarifier. The total dry mass of solids produced is most nearly

- (A) 6000 kg/d
- (B) 7000 kg/d
- (C) 9000 kg/d
- (D) 11 000 kg/d

47. Anaerobic digesters receive a total of 3000 m^3/d of primary sludge at an ultimate BOD concentration of 400 mg/L. The digested sludge is wasted at a rate of 30 m^3/d with a suspended solids concentration of 10 000 mg/L. The efficiency of waste utilization is 0.7. The volume of methane produced is most nearly

- (A) 140 m^3/d
- (B) 190 m^3/d
- (C) 220 m^3/d
- (D) 260 m^3/d

48. Sulfur dioxide is to be used to dechlorinate an effluent containing a chlorine residual of 6 mg/L as Cl_2. The design flow is 80 L/s. The amount of sulfur dioxide required is most nearly

- (A) 10 kg/d
- (B) 20 kg/d
- (C) 30 kg/d
- (D) 40 kg/d

49. An anoxic basin will be used to denitrify a wastewater with the characteristics given in the following table.

parameter	value
influent NO_3-N	26 mg/L
effluent NO_3-N	4 mg/L
MLVSS	2500 mg/L
DO	0.2 mg/L
temperature	10°C
specific denitrification rate	0.09 kg NO_3-N/ kg MLVSS·d

The required detention time is most nearly

- (A) 5 h
- (B) 6 h
- (C) 7 h
- (D) 8 h

50. A plating factory has a discharge permit that requires effluent to be diluted at least 20 to 1 in the nearfield for a maximum flow of 75 ft^3/sec into a river. At low river flow, the diffuser is submerged 4 ft and the stream velocity is 3.0 ft/sec. The diffuser is 2 ft above the river bed. The diffuser has 10 ports, each with a 12 in diameter, discharging 60° above the horizontal. What is most nearly the required diffuser length?

- (A) 20 ft
- (B) 40 ft
- (C) 60 ft
- (D) 80 ft

51. A constructed wetlands design is being reviewed. The design parameters are given in the following table.

parameter	value
length of basin	200 ft
width of basin	2500 ft
fraction of cross section not occupied by plants	0.75
depth of basin	2 ft
average flow rate	40,000 ft^3/day
specific surface area for microbiological activity	4.8 ft^2/ft^3
influent BOD_5	200 mg/L
fraction of BOD_5 not removed by settling at head of system	0.52
rate constant at 20°C	0.006/day
temperature	10°C

The effluent BOD_5 concentration will be most nearly

- (A) 40 mg/L
- (B) 65 mg/L
- (C) 80 mg/L
- (D) 95 mg/L

52. Wastewater containing a fine particulate is to be polished with a dissolved air flotation unit. The design criteria are given in the following table.

parameter	value
suspended solids	300 mg/L
optimum flotation air-to-solids ratio	0.07 mg air/mg solids
operating temperature	20°C
air solubility at 20°C	18.7 mL air/L water
absorption fraction	0.6

For pressurization of the total wastewater flow, the required gage pressure is most nearly

(A) 160 kPa
(B) 180 kPa
(C) 200 kPa
(D) 220 kPa

53. Which of the following is NOT a cause of high concentrations of suspended solids in wastewater effluent?

(A) inadequate solids removal, causing nitrogen bubbles and rising sludge in secondary clarifiers
(B) out of balance food-to-microorganism ratio, causing bulking sludge
(C) insufficient conditioning chemical for sludge dewatering, causing cycling of fine solids
(D) excessive activated sludge return rates, causing dilute return activated sludge

54. BOD analyses of an industrial wastewater give the results shown in the following table. Assume the samples are places in standard 300 mL BOD bottles.

bottle no.	wastewater portion (mL)	initial DO (mg/L)	final DO (mg/L)
1	3	8.6	6.0
2	3	8.6	6.1
3	3	8.4	6.6
4	2	8.5	6.1
5	2	8.5	6.4
6	2	8.6	6.0
7	1	8.6	6.8
8	1	8.5	6.4
9	1	8.4	6.5

Based on these analyses, which is the principal conclusion that may be drawn?

(A) The wastewater shows a high BOD, thus confirming an industrial waste.
(B) The BOD results vary widely, indicating an inaccuracy in the test.
(C) The high BOD results indicate the need to run a wider range of dilutions.
(D) The increasing BOD with increasing dilution indicates toxicity in the wastewater.

55. Which of the following statements are true?

I. The abundance of algae in natural waters depends on the availability of nitrogen and phosphorus.

II. Algae in oxidation ponds can interfere with water treatment and so should be removed regularly.

III. Only two constituents, nitrogen and carbon, need to be controlled in wastewater effluent to meet water quality objectives related to algae.

IV. The use of carbon dioxide by algae may cause high diurnal variations in pH in stabilization ponds.

V. Control of algae in natural waters that serve as water supply is important due to the algae's potential to cause taste and odor problems.

(A) I, III, IV
(B) I, IV, V
(C) II, III, V
(D) III, IV, V

56. The removal of coliform organisms in a small stream is analyzed. The approximate initial die-away rate of the bacteria population is 1500/h. The coefficient of nonuniformity or retardation is 6.15. The stream has a flow of 400 L/s, a depth of 1 m, and a width of 10 m. What is most nearly the percent removal 10 km downstream from the point where the analysis took place?

(A) 30%
(B) 50%
(C) 70%
(D) 90%

57. Which of the following statements are true relative to the effects on stream biology from organic loads?

I. Nitrogen in effluent from conventional wastewater treatment plants does not add to the organic load because it is normally not biologically degradable.

II. Effluents that are disinfected do not add organic load to a stream because bacterial populations are too low to promote decomposition and oxygen demand.

III. As dissolved oxygen sags from an organic load, the variety of biological life decreases and the numbers within the surviving species rise rapidly.

IV. Immediately below the outfall of partially stabilized wastewater, the populations of algae decrease; but as nutrient salts are assimilated, the algae greatly increase in numbers.

V. Ammonia in wastewater effluents can be toxic to aquatic organisms, so its concentration in the stream must be reduced by dilution or controlled by advanced treatment.

(A) I, III, V
(B) I, II, IV
(C) II, IV, V
(D) III, IV, V

58. Between the following lists, match the food chain elements or life forms with the appropriate requirements or responses.

food chain elements or life form

I. aquatic life diversity and abundance

II. algae and green plants

III. fly nymphs, copepods, and water fleas

IV. sunfish

V. bass, pike, and salmon

VI. decomposed fish and other aquatic life

food chain requirement or response

1. is (are) first-order consumers in the food chain

2. can be limited by the absence of sunlight or the presence of pollution

3. is (are) second-order consumers of small herbivores

4. release(s) nutrients that are recycled into algae by photosynthesis

5. require(s) the presence of oxygen, carbon dioxide, nitrogen, and phosphorus

6. is (are) third-order consumers of flesh eaters

7. is (are) primary producers since they use the energy of sunlight to synthesize inorganic substances into living tissue

Which option represents the correct matches?

(A) I-2, I-5, II-7, III-1, IV-3, V-6, VI-4
(B) I-7, II-1, III-3, IV-6, V-7, VI-2
(C) I-1, II-1, III-6, IV-2, V-1, VI-5
(D) I-4, II-3, III-7, IV-6, V-3, VI-2

59. A pulp mill discharges a treated effluent to a river where complete mixing occurs quickly below the outfall. The effluent and stream conditions are given in the following table.

parameter	value
effluent flow	$0.5 \text{ m}^3/\text{s}$
effluent BOD_5 at 20°C	60 mg/L
effluent DO	2 mg/L
effluent temperature	25°C
deoxygenation rate, K_d, base 10	0.15 d^{-1}
temperature variation constant θ_d for K_d	1.046
stream flow	$6 \text{ m}^3/\text{s}$
stream velocity	0.3 m/s
stream BOD_5 at 20°C before mix	4.5 mg/L
stream DO before mix	8.5 mg/L
stream temperature before mix	15°C
reaeration rate, K_r, base 10	0.250 d^{-1}
temperature variation constant θ_r for K_r	1.024

The DO of the stream at the critical point is most nearly

(A) 4.8 mg/L
(B) 5.7 mg/L
(C) 6.8 mg/L
(D) 7.6 mg/L

60. Which of the following statements regarding eutrophication of lakes or impoundments are true?

I. A common indicator of eutrophic waters is the abundance of blue-green algae as compared to other algae species.

II. Eutrophic effects can be rapidly overcome by limiting nitrogen and phosphorus in wastewater discharges.

III. Phosphorus has been identified as more critical than nitrogen in controlling algae and aquatic weeds.

IV. The most common methods of controlling eutrophication caused by point sources are diversion of discharges to a different drainage or treatment to remove nitrogen and phosphorus.

V. Bottom sediments rarely contribute to euthrophic effects.

(A) I, II, IV
(B) I, III, IV
(C) II, III, IV
(D) II, IV, V

61. The first stage BOD of a wastewater is 150 mg/L at 20°C and a rate constant of 0.23/d (base e). If the same waste concentration is discharged at a temperature of 30°C, the BOD_5 will be most nearly

(A) 100 mg/L
(B) 125 mg/L
(C) 150 mg/L
(D) 175 mg/L

62. Which of the following statements are true concerning indicator organisms?

I. Coliform organisms are a useful indicator of the potential contamination of water supplies or the degree of pollution of natural waters because they are present in large numbers in sewage.

II. A distinguishing characteristic of coliforms is their ability to ferment and produce carbon dioxide gas in 48 hr.

III. The membrane filter technique provides only a qualitative presumptive test for coliforms that must be confirmed by multiple-tube fermentation.

IV. *Aerobacter aerogenes* is not a practical indicator organism because it is commonly found in soil.

V. A characteristic that makes coliform organisms good indicators is their abundance in human wastes.

(A) I, II, III
(B) I, II, V
(C) I, III
(D) II, III, IV

63. A wastewater treatment plant is required to produce an effluent with a coliform count of less than 200/100 mL. Before it is disinfected, the wastewater is found to average 2×10^8/100 mL coliform at a peak hourly flow of 400 L/s. The chlorine contact tank has an effective volume of 1440 m^3. The chlorine residual required to meet the effluent limitation is most nearly

(A) 4.3 mg/L
(B) 5.7 mg/L
(C) 6.5 mg/L
(D) 7.2 mg/L

64. Which of the following may be associated with tastes and odors in rivers, lakes, and other fresh waters?

I. *Ceriodaphnia dubia* (daphnid shrimp)

II. *Synura* (flagellate algae)

III. *Anabaena* (blue-green algae)

IV. *Pimephales promelas* (fathead minnow)

V. *Oscillatoria* (blue-green algae)

VI. *Streptomyces* (mold-like filamentous bacteria)

VII. *Bacillus subtilis* (spore-forming bacteria)

(A) I, II, VI, VII
(B) II, III, V, VI
(C) III, IV, V, VII
(D) IV, V, VI, VII

65. An MPN test gave the results shown in the following table.

serial dilution	sample portion (mL)	no. of positive reactions out of five tubes
0	1.0	5
1	0.1	5
2	0.01	5
3	0.001	2
4	0.0001	1
5	0.00001	0

Using MPN tables, the MPN is

(A) 70/100 mL
(B) 700/100 mL
(C) 7000/100 mL
(D) 70 000/100 mL

66. BOD test results for raw domestic settled wastewater are given in the following table.

dilution no.	wastewater volume (mL)	initial DO (mg/L)	5 d DO (mg/L)
1	5	8.0	6.2
2	10	8.2	5.2
3	15	8.4	3.5

For a deoxygenation rate constant (base *e*) of 0.25/d, the ultimate BOD is most nearly

(A) 120 mg/L
(B) 130 mg/L
(C) 140 mg/L
(D) 150 mg/L

67. Which of the following actions can improve the biological quality of fresh water?

I. control the concentration of nitrogen and phosphorus to limit algae growth

II. apply copper sulfate periodically to limit algae growth

III. change the hydrography to decrease stream velocity

IV. add chlorine to kill pathogenic bacteria

V. control wastewater discharges to limit DO depletion

   (A) I, II
   (B) I, V
   (C) III, IV
   (D) IV, V

68. Recyclable solid wastes are source separated at the residences and hauled to a recycling center for processing. The types of waste separated for every 100 lbm of total waste generated are given in the following table.

component	total waste (lbm)	recyclable wastes separated (lbm)	as-collected density (lbm/ft^3)
paper	36	21	6
cardboard	6	4	3
food wastes	8	–	18
plastic	7	2	4
yard debris	15	–	6
wood	2	–	14
glass	8	3	12
tin cans	6	1	5
all other	12	–	10
total	100	31	

A total of 2000 residences are served with an average of 3.5 persons per residence and an average waste of 4 lbm/person-day. The participation rate is 80% for paper and cardboard and 50% for other categories of recyclables. If a 15 yd^3 collection vehicle is used at 90% efficiency in volume utilization, the number of trips required per week is most nearly

   (A) 12
   (B) 16
   (C) 24
   (D) 32

69. A transfer station must serve both packer trucks and small vehicles with design characteristics given in the following table.

parameter	value packer trucks	small vehicles
average payload	6 U.S. tons	0.4 U.S. tons
peak month/ average month	1.5	1.0
peak hour/ average hour	2.0	1.0
unloading time	6 min	15 min

The transfer station receives an average of 500 U.S. tons/day with 80% from collection vehicles and 20% from small vehicles. What is most nearly the minimum number of unloading bays to accommodate the peak hour without waiting?

   (A) 7
   (B) 11
   (C) 18
   (D) 23

70. A processed solid waste has a basic composition that can be approximated as $C_{40}H_{55}O_{30}$. A pilot test estimated that partial aerobic conversion would result in residual organic matter with an approximate composition of $C_{12}H_{25}O_{10}$. The initial mass of solid waste is 1000 kg, and the mass of the residual is 350 kg. The aerobic stabilization process is described by

$$C_aH_bO_cN_d + 0.5(ny + 2s + r - c)O_2$$
$$\rightarrow nC_wH_xO_yN_z + sCO_2 + (d - nz)NH_3$$
$$r = 0.5\big(b - nx - 3(d - nz)\big)$$
$$s = a - nw$$

The mass of oxygen required for the conversion is most nearly

   (A) 450 kg
   (B) 530 kg
   (C) 680 kg
   (D) 770 kg

71. A landfill has solid waste cells at a compacted density of 1000 lbm/yd^3 and uses soil cover at an in-place density of 3000 lbm/yd^3. The soil-to-solid waste ratio is 1:5. No moisture is added. For a solid waste moisture content of 20% and a 12 ft lift, the mass of leachate generated is most nearly

   (A) 0 lbm
   (B) 250 lbm
   (C) 500 lbm
   (D) 750 lbm

72. Solid waste from a small community has the composition given in the following table.

component	component proportions in the waste stream as collected with no recycling (% by weight)
food waste	10
paper	35
cardboard	5
plastics	8
textiles	2
yard waste	20
glass	8
all other	12
total	100

The community is considering implementing a recycling program for paper and cardboard with expected effectiveness of 70% and 80%, respectively. If the program achieves the expected effectiveness, the percent by weight of yard waste in the waste stream would be most nearly

(A) 23%
(B) 28%
(C) 34%
(D) 36%

73. A solid waste transfer proposal is being compared to the existing direct haul operation. The characteristics of the two systems are given in the following table.

parameter	value compactor truck direct haul	value semi-trailer transfer
capacity in volume	15 yd³	100 yd³
transfer station operation cost	–	$3/yd³
operation cost	$40/hr	$50/hr

The minimum round trip time for which the semi-trailer system would be more economical is most nearly

(A) 85 min
(B) 90 min
(C) 95 min
(D) 100 min

74. A mixed solid waste, mass-fired energy recovery plant—comprised of a steam boiler, turbine, and generator—has the design values given in the table.

parameter	value
mixed solid waste throughput	1500 U.S. tons/day
energy value of mixed solid waste	4800 Btu/lbm
station service allowance of total power produced	4%
unaccounted-for heat loss of total power produced	6%
boiler efficiency	72%
turbine efficiency	28%
electric generator efficiency	92%

The overall efficiency of the plant is most nearly

(A) 9%
(B) 12%
(C) 17%
(D) 19%

75. A community water supply is contaminated with MTBE and requires treatment to protect residents. The reference dose oral route for MTBE is 0.005 mg/kg·d. An adult has a body mass of 70 kg and an average daily intake of 2 L. For 120 d/yr exposure and complete absorption, the maximum water concentration to protect an adult is most nearly

(A) 0.03 mg/L
(B) 0.36 mg/L
(C) 0.53 mg/L
(D) 1.3 mg/L

76. Which of the following statements are true regarding solid/hazardous waste standards?

I. RCRA Subtitle C addresses nonhazardous waste, and RCRA Subtitle D addresses hazardous wastes.

II. Landfills with a design capacity of 2.5×10^9 kg or greater are subject to USEPA landfill gas emission regulations.

III. USEPA drinking water standards for maximum contaminant levels (MCLs) apply to groundwater protection from sanitary landfills only beyond the property boundary.

IV. USEPA criteria for sanitary landfills require installation of a composite liner system to intercept, collect, and remove any leachate that migrates from the landfill.

V. USEPA criteria require groundwater monitoring for a sanitary landfill.

(A) I, II, V
(B) I, III, V
(C) II, III, IV
(D) II, IV, V

77. An infiltration gallery intercepts groundwater and uses the collected water for irrigation. The stratum is 6 m thick and consists of clean sand and gravel with a coefficient of permeability of 0.15 cm/s. The diffusion ditch to the gallery penetrates to the sole of the stratum. The water surface in the gallery is 0.5 m above the sole of the stratum, and the gallery is 10 m horizontally from the diffusion ditch. The flow into the gallery per meter of length is most nearly

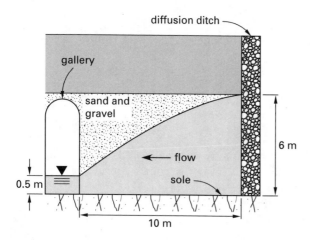

(A) 0.5×10^{-3} m^3/m·s
(B) 1×10^{-3} m^3/m·s
(C) 2×10^{-3} m^3/m·s
(D) 3×10^{-3} m^3/m·s

78. Pumping tests for a homogeneous aquifer are conducted as shown in the following table. Assume the drawdown compared to aquifer thickness is small, and the length of time of the pumping test is relatively long.

parameter	value
duration of pumping test	3 d
pumping rate	50 L/s
well diameter	450 mm
radius of influence	300 m
elevation of top of aquifer	175 m
elevation of bottom of aquifer	55 m
water surface elevation	
at 300 m from well	100 m
drawdown in well	3.5 m
elevation of well screen	55–100 m

The coefficient of permeability of the aquifer is most nearly

(A) 200×10^{-6} m/s
(B) 400×10^{-6} m/s
(C) 600×10^{-6} m/s
(D) 800×10^{-6} m/s

79. A groundwater has the chemical analysis given in the following table.

constituent	concentration
Ca^{+2}	90 mg/L
Mg^{+2}	38 mg/L
Na^+	9.7 mg/L
K^+	5.2 mg/L
Fe^{+2}	0.10 mg/L
HCO_3^-	383 mg/L
SO_4^{-2}	39 mg/L
Cl^-	36 mg/L
F^-	0.2 mg/L
NO_3^-	2.2 mg/L
SiO_2	16 mg/L
total dissolved solids	432 mg/L
pH	7.0

The total hardness as $CaCO_3$ of the water is most nearly

(A) 380 mg/L
(B) 420 mg/L
(C) 460 mg/L
(D) 500 mg/L

80. A soil of great depth and approximately uniform permeability is to be drained with drains uniformly spaced with inverts at 3.5 m below the water table. The water level in the drains will be at the crown of the pipe. The average permeability of the sandy soil is 350 m/yr, and groundwater will flow radially to the drains. For a drain length of 1200 m and a drain spacing of 100 m, the total flow from each drain is most nearly

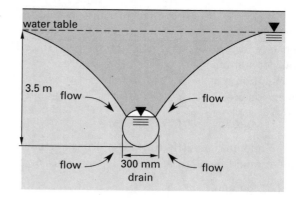

(A) 0.015 m^3/s
(B) 0.020 m^3/s
(C) 0.025 m^3/s
(D) 0.030 m^3/s

Afternoon Session
Geotechnical

81. A rigid foundation is supported by friction piles in clay as shown in the following plan and elevation views. The total load on the piles, reduced by the displaced soil weight, is 2000 kN. The settlement of layer 2 is most nearly

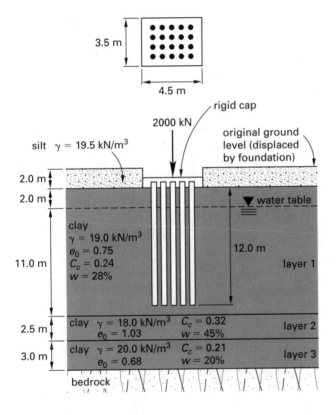

(A) 12 mm
(B) 17 mm
(C) 24 mm
(D) 35 mm

82. A 250 mm layer of soil bentonite will be placed just beneath the geomembrane liner of a proposed landfill. The soil bentonite layer will be placed in two 125 mm lifts, and the bentonite content will be 8% by dry weight. If the compacted moist unit weight of this soil bentonite layer is 17.0 kN/m^3 with a moisture content of 18%, the amount of dry bentonite that must be spread for mixing on each lift is most nearly

(A) 0.14 kN/m^2
(B) 0.18 kN/m^2
(C) 0.22 kN/m^2
(D) 0.26 kN/m^2

83. A landfill is 300 m × 400 m in plan. The clay liner has a hydraulic conductivity of 6×10^{-7} cm/s and experiences an average annual leachate head of 0.5 m as shown. A subgrade drain (pore pressure is atmospheric) lies below the clay liner. Assume one-dimensional flow downward for the leachate. The annual steady-state flow rate from this landfill is most nearly

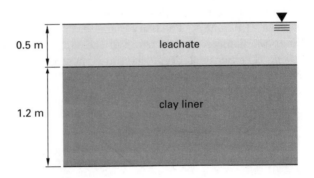

(A) 10^3 m^3/year
(B) 10^4 m^3/year
(C) 10^5 m^3/year
(D) 10^6 m^3/year

84. The groundwater table on a project site (elevation view shown) will be lowered 16.3 m. The groundwater table is now at the ground surface. Assume a soil moisture content of 11% above the groundwater table once it is lowered. After lowering, the settlement of the clay layer will be most nearly

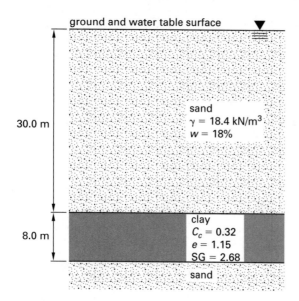

(A) 0.10 m
(B) 0.20 m
(C) 0.30 m
(D) 0.40 m

85. The permeability of a soil is evaluated in a falling-head permeameter. The head decreases from 100 cm to 50 cm in 21 min 18 s. The body diameter is 10 cm, the standpipe diameter is 0.25 cm, and the sample length is 6 cm. The permeability of the soil is most nearly

(A) 1×10^{-7} cm/s
(B) 2×10^{-7} cm/s
(C) 1×10^{-6} cm/s
(D) 2×10^{-6} cm/s

86. A site consists of 25 m of clayey silt that is to be consolidated for eventual placement of a large office building. From a consolidation test with a soil sample 5.0 cm high, it has been determined that the time to achieve 90% consolidation (of the soil sample) is 10 min 46 s. Assuming double drainage for both the sample and the clayey silt layer, how much time would be required to achieve 90% consolidation of the 25 m clayey silt layer?

(A) 1 yr
(B) 2 yr
(C) 5 yr
(D) 15 yr

87. A clay layer 10 m thick (with double drainage) is expected to have an ultimate settlement of 502 mm. If the settlement in 5 yr is 124 mm, the remaining time it will take to reach a settlement of 250 mm is most nearly

(A) 5 yr
(B) 10 yr
(C) 15 yr
(D) 20 yr

88. A smooth retaining wall holds back a uniform sand backfill as shown. The distance of the active resultant force from the bottom of the retaining wall is most nearly

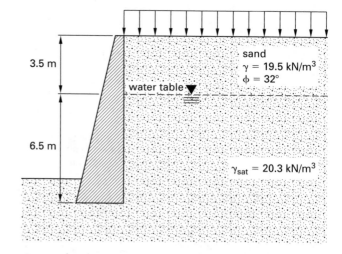

(A) 2.9 m
(B) 3.3 m
(C) 3.8 m
(D) 4.5 m

89. A loose, natural sand deposit has a saturated unit weight of 19.3 kN/m^3 and an angle of internal friction of 29°. The water table is at the ground surface. The total at-rest lateral earth pressure at a depth of 10 m is most nearly

(A) 80 kPa
(B) 150 kPa
(C) 210 kPa
(D) 240 kPa

90. An artificial reservoir holds a constant level of water as shown. A compacted clay liner with the given properties is used to contain the water. The true water velocity (pore velocity) through the clay liner is most nearly

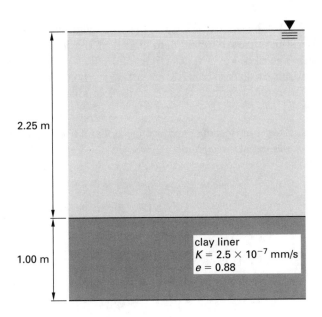

(A) 5.6×10^{-7} mm/s
(B) 6.7×10^{-7} mm/s
(C) 1.2×10^{-6} mm/s
(D) 3.1×10^{-6} mm/s

91. A sample of saturated clay has a total mass of 1733 g and a dry mass of 1287 g. The specific gravity of the soil particles is 2.7. The total unit weight of this soil is most nearly

(A) 17.1 kN/m^3
(B) 17.7 kN/m^3
(C) 18.0 kN/m^3
(D) 18.4 kN/m^3

92. A smooth gravity retaining wall holds soil backfill with properties as shown. Disregard passive earth pressure. The vertical pressure at point A is most nearly

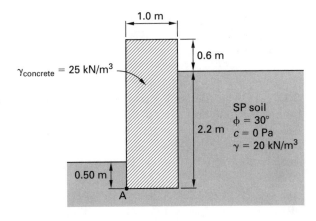

(A) 100 kPa
(B) 120 kPa
(C) 125 kPa
(D) 140 kPa

93. A medium uniform sand has the gradation shown. The sand has a dry unit weight of 15.8 kN/m^3, and the particles have a specific gravity of 2.65.

sieve no.	sieve size (mm)	percent finer (by mass)
10	2.00	100.0
20	0.850	99.7
30	0.600	93.0
40	0.425	58.2
50	0.300	42.9
70	0.212	18.2
100	0.150	10.1
200	0.075	1.0

The estimated coefficient of permeability for this sand is most nearly

(A) 1.0×10^{-3} cm/s
(B) 3.0×10^{-3} cm/s
(C) 6.0×10^{-3} cm/s
(D) 4.0×10^{-2} cm/s

94. The soil profile and the properties of each soil layer beneath a reservoir are shown. The sandy layer at the bottom of the soil profile has horizontal drainage and zero pore pressure. The water level of the reservoir is constant, and the total area of the reservoir is 5000 m^2. Assuming vertical flow through the soil profile, the water loss from the reservoir in 6 mo is most nearly

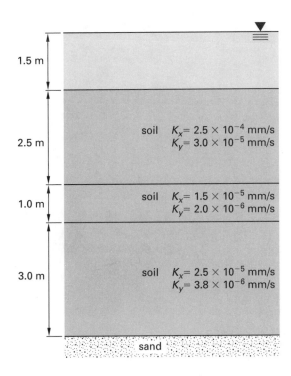

(A) 85 m³
(B) 94 m³
(C) 1000 m³
(D) 1200 m³

95. A concrete dam impounds water. Using the flow net shown, the pore pressure at point A is most nearly

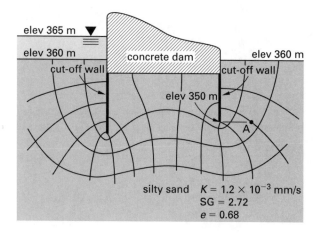

(A) 80 kPa
(B) 105 kPa
(C) 125 kPa
(D) 140 kPa

96. What is the effective area of the rectangular footing supporting a concentrated normal force as shown?

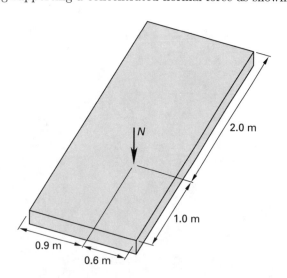

(A) 0.6 m²
(B) 1.8 m²
(C) 2.4 m²
(D) 4.5 m²

97. A long wall footing that is 2 m wide is situated on stiff, saturated clay. The depth of the footing is 1 m. The clay has a unit weight of 18.5 kN/m³ and an undrained shear strength of 110 kPa. Loading is applied rapidly enough that undrained conditions prevail ($\phi = 0$).

Use Terzaghi bearing capacity factors and the following bearing capacity formula.

$$q_{\text{ult}} = c\lambda_{cs}\lambda_{cd}N_c + q\lambda_{qs}\lambda_{qd}N_q + \tfrac{1}{2}\lambda_{\gamma s}\lambda_{\gamma d}\gamma BN_\gamma$$

The shape and depth factors are

$$\lambda_{qs} = \lambda_{\gamma s} = 1$$
$$\lambda_{qd} = \lambda_{\gamma d} = 1$$
$$\lambda_{cs} = 1 + 0.2\frac{B}{L}\tan^2\left(45 + \frac{\phi}{2}\right)$$
$$\lambda_{cd} = 1 + 0.2\frac{D_f}{B}\tan\left(45 + \frac{\phi}{2}\right)$$

The ultimate bearing capacity per meter length of footing is most nearly

(A) 300 kN/m
(B) 600 kN/m
(C) 1000 kN/m
(D) 1400 kN/m

98. A rock core is retrieved from a drill hole. The length of the recovered core is 123 cm. There are five pieces 10 cm or more in length, and the pieces have a combined length of 89 cm.

The rock quality designation for this core is most nearly

(A) 0%
(B) 38%
(C) 72%
(D) 138%

99. A 0.30 m diameter prestressed concrete pile has been driven 6 m into a dense sand deposit. The soil-pile friction angle is 25°. The unit weight of the prestressed concrete pile is 25 kN/m³, and the unit weight of the sand is 20 kN/m³. Assume that the critical depth is 20 times the diameter of the pile, and that the horizontal earth pressure coefficient for tension is 1.1. The ultimate pullout load capacity of the pile is most nearly

(A) 160 kN
(B) 170 kN
(C) 180 kN
(D) 190 kN

100. A soil profile has the properties shown. The average permanent vertical pressure on the normally consolidated clay layer is expected to increase by 130 kPa. The average effective overburden pressure at the middle of the clay layer is 240 kPa. The total primary consolidation settlement is most nearly

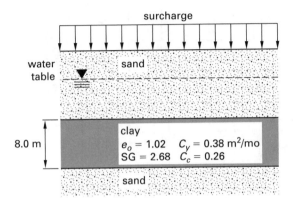

(A) 100 mm
(B) 180 mm
(C) 190 mm
(D) 200 mm

101. The clay soil shown undergoes consolidation. The percent consolidation at mid-depth of the clay 3 yr after loading is most nearly

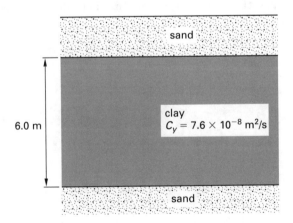

(A) 30%
(B) 40%
(C) 70%
(D) 90%

102. A double-drained clay layer 20 m thick settles 18.2 cm in 5 yr. The coefficient of consolidation for this clay is 4.3×10^{-7} m^2/s. The time required for the clay

layer to undergo 90% of its ultimate primary consolidation settlement amount is most nearly

(A) 4.2 yr
(B) 5.8 yr
(C) 6.3 yr
(D) 7.2 yr

103. A clay soil is loaded as shown and undergoes consolidation. If one-dimensional loading is assumed, what is most nearly the excess pore water pressure at mid-depth of the clay layer immediately after loading?

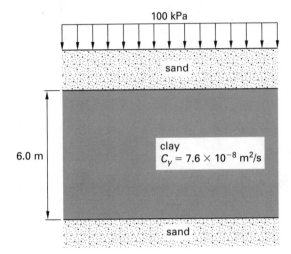

(A) 0 kPa
(B) 35 kPa
(C) 71 kPa
(D) 100 kPa

104. A soil has a wet unit weight of 17.6 kN/m^3 and a moisture content of 8%. The specific gravity of the solid particles is 2.72. The degree of saturation is most nearly

(A) 8%
(B) 14%
(C) 28%
(D) 34%

105. A soil has a grain size distribution as shown in the sieve analysis chart. (See accompanying chart that follows.) The coefficient of curvature for this soil is most nearly

(A) 2.0
(B) 3.5
(C) 430
(D) 450

Chart for Problem 105

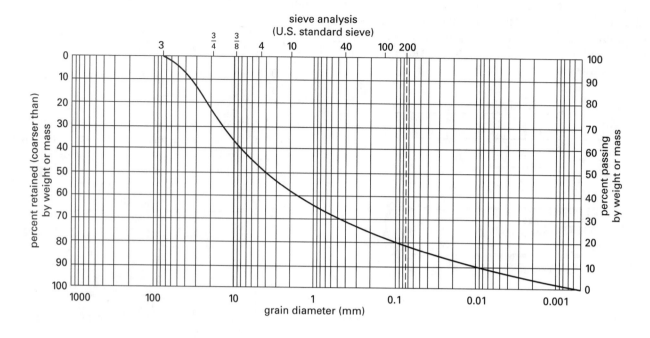

106. A braced cut in clay has properties and dimensions as shown in the illustration. The horizontal center-to-center spacing of the struts is 4.0 m. The load on the bottom strut (strut A) is most nearly

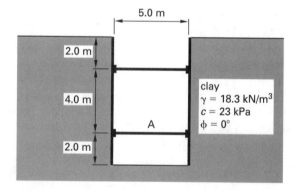

(A) 190 kN
(B) 440 kN
(C) 940 kN
(D) 1000 kN

107. A 10 m long precast concrete pile is installed into homogeneous sand. The pile cross section is 254 mm × 254 mm. The unit weight of the sand is 18.8 kN/m³. The internal friction angle of the sand is 35°. Consider the critical depth as 15 times the width of the pile. If the earth pressure coefficient is 1.6 and the soil-pile friction angle is 0.6ϕ, then the total frictional resistance of the pile in the sand is most nearly

(A) 80 kN
(B) 250 kN
(C) 360 kN
(D) 480 kN

108. A wall is supporting a horizontal force as shown. Assume that the wall is smooth. The total force per meter of wall that the soil can sustain is most nearly

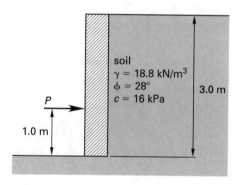

(A) 150 kN/m
(B) 395 kN/m
(C) 480 kN/m
(D) 495 kN/m

109. A consolidated-drained test (S-test) is performed on a sand sample. Initially, the saturated sand is consolidated in the triaxial cell under an equal all-around pressure of 200 kPa. Maintaining the cell pressure, the axial stress is increased 468 kPa. Under this stress state, the sample is at failure. The angle of internal friction of the sample is most nearly

(A) 0°
(B) 30°
(C) 33°
(D) 38°

110. A layer of sand has particles with a specific gravity of 2.66 and a void ratio of 0.62. The buoyant unit weight of the sand is most nearly

(A) 10 kN/m^3
(B) 11 kN/m^3
(C) 12 kN/m^3
(D) 20 kN/m^3

111. A continuous wall footing 1.5 m wide supports a load of 596 kN/m. The unit weight of the soil beneath the foundation is 18.6 kN/m^3. The soil has a cohesion of 14 kPa and an angle of internal friction of 25°. If the footing is placed near the ground surface, and if the Terzaghi bearing capacity factors and formula are used, the factor of safety against bearing capacity failure is most nearly

(A) 0.8
(B) 1.2
(C) 1.8
(D) 2.8

112. A dry sand sample is tested in a direct shear box with a normal stress of 100 kPa. Failure occurs at a shear stress of 63.4 kPa. The size of the tested sample is 6 cm × 6 cm × 3 cm (height). For a normal stress of 75 kPa, what shear force would be required to cause failure in the sample?

(A) 0.17 kN
(B) 0.37 kN
(C) 2.8 kN
(D) 48 kN

113. An unconfined-undrained compression test is conducted on a clay soil sample that had an initial height of 9.1 cm and an initial diameter of 4.0 cm. The axial load at failure is 0.43 kN, and the corresponding height is 8.67 cm. The undrained shear strength of this clay is most nearly

(A) 80 kPa
(B) 160 kPa
(C) 180 kPa
(D) 320 kPa

114. A consolidated-drained test is performed on a normally consolidated clay. The chamber confining pressure is 280 kPa, and the deviator stress at failure is 410 kPa. Assume that the normally consolidated clay has no drained cohesion ($c' = 0$). The shear stress on the failure plane is most nearly

(A) 190 kPa
(B) 300 kPa
(C) 380 kPa
(D) 580 kPa

115. A sand has a minimum void ratio of 0.41 and a maximum void ratio of 0.78. Its dry unit weight is 16.5 kN/m^3. If the specific gravity of the solids is 2.65, the relative density of this sand is most nearly

(A) 0.40
(B) 0.55
(C) 0.65
(D) 0.80

116. A soil has the following properties.

liquid limit	40
plasticity index	13
percent passing no. 10 sieve	98%
percent passing no. 40 sieve	87%
percent passing no. 200 sieve	45%

The AASHTO classification and group index number is most nearly

(A) A-5 (3)
(B) A-6 (1)
(C) A-6 (3)
(D) A-7-6 (1)

117. Classify a soil with the following characteristics using the Unified Soil Classification System (USCS).

liquid limit	55
plastic limit	20
C_u	12
C_c	1.5
percent passing 1 in sieve	100%
percent passing no. 4 sieve	98%
percent passing no. 40 sieve	45%
percent passing no. 200 sieve	26%

(A) SC
(B) SW
(C) SP
(D) SM

118. The active pressure on the sheet pile wall shown is in equilibrium with the passive pressure and the anchor force. Assume that the sheet pile is smooth and that the resultant force acting on the passive side is horizontal and acting at point C as shown. The value of the force per meter length of wall is most nearly

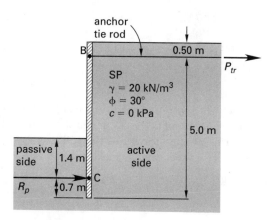

(A) 70 kN/m
(B) 80 kN/m
(C) 100 kN/m
(D) 120 kN/m

119. A mat foundation is 20 m × 31 m in plan. The total dead and live load is 33 540 kN. The depth needed for a fully compensated foundation is most nearly

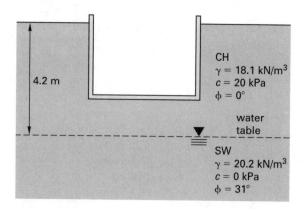

(A) 0.80 m
(B) 2.0 m
(C) 3.0 m
(D) 3.5 m

120. A concrete retaining wall has the specifications shown. If passive resistance is ignored, the factor of safety against sliding is most nearly

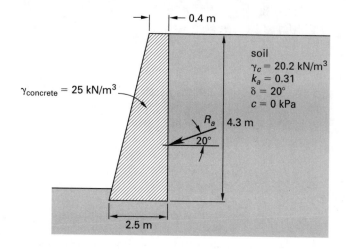

(A) 1.0
(B) 1.2
(C) 1.6
(D) 2.8

Afternoon Session
Structural

121. A two-lane highway bridge is constructed using precast concrete girders spaced 7 ft on center. The girders are simply supported and span 60 ft. The weight of girders and deck is such that the dead load bending moment at the critical location for bending moment is 500 ft-kips. If the bridge is designed for AASHTO HS20-44 loading using load factor design, the design bending moment at the critical location is most nearly

(A) 400 ft-kips
(B) 1000 ft-kips
(C) 1800 ft-kips
(D) 2100 ft-kips

122. A rigid diaphragm transfers a lateral wind force of 0.4 kips/ft into a system of shearwalls whose relative rigidities against forces in the north direction are shown in the plan. The force in wall A of the system is most nearly

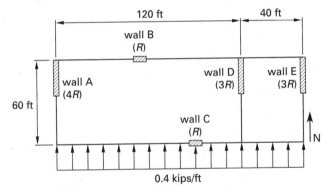

(A) 15 kips
(B) 22 kips
(C) 27 kips
(D) 33 kips

123. The roof framing of a single story commercial building consists of wood joists supported by timber beams and sheathed with a properly nailed and blocked plywood diaphragm. Seismic lateral forces for NS ground motion are shown. Assume sufficiently rigid plywood shearwalls 14 ft high and 24 ft long are constructed at lines 1, 2, and 3. The axial compression and tension forces in the shearwall boundary members at line 2 under the given loadings are most nearly

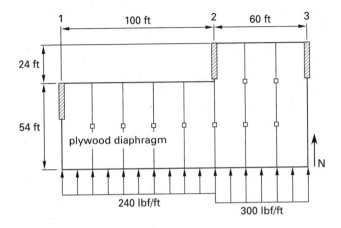

(A) 10 kips
(B) 12 kips
(C) 16 kips
(D) 20 kips

124. A two-story building is 14 ft from ground to second floor and 12 ft from second floor to roof. The exterior wall projects 3 ft above the roof level to create a parapet. The exterior wall weighs 15 psf, the second floor dead load is 30 psf, and the roof dead load is 20 psf. The building is wood framed with plywood diaphragms and shearwalls resisting all lateral forces. The building is situated in seismic performance category D, where the design spectral response acceleration at short periods is 0.6, the design spectral response acceleration at one second period is 0.2, and the importance factor for seismic response is 1.0. Assume the building qualifies as a building frame system with light-frame walls with shear panels. The seismic base shear for NS ground motion by the static force procedure of the IBC, on a working load basis, is most nearly

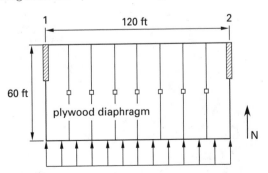

(A) 25 kips
(B) 45 kips
(C) 65 kips
(D) 80 kips

125. An 8 in thick bridge deck is made of reinforced normal weight concrete. It is supported by steel girders spaced 8.5 ft on center, with flange widths of 1 ft. The bending moment, per foot of width, for dead weight of the slab is 1 ft-kip/ft at the critical location. If the bridge is to be designed using load factor design for an AASHTO HS 20 loading, and the slab is continuous over three or more spans, the design bending moment per foot width of slab is most nearly

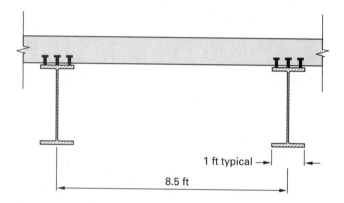

(A) 6 ft-kips/ft
(B) 8 ft-kips/ft
(C) 10 ft-kips/ft
(D) 12 ft-kips/ft

126. The circular shaft shown is subjected to an axial tension force P at its free end and a compressive force of 50 kips at point B. Note that the shaft is hollow between points A and B. The allowable normal tension stress is 22 ksi, the modulus of elasticity is 29,000 ksi, and the maximum allowable elongation is 0.04 in. The maximum allowable value of P is most nearly

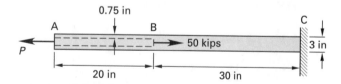

(A) 90 kips
(B) 120 kips
(C) 150 kips
(D) 170 kips

127. The compound beam shown has an internal hinge ($M = 0$) at point B and is simply supported on hinges or rollers at points A, C, and E. The ordinate of the influence line for the bending moment at point D, which is 12 ft to the right of support C, is most nearly

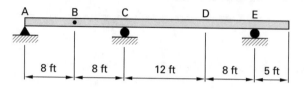

(A) 1 ft-kips/kip
(B) 3 ft-kips/kip
(C) 5 ft-kips/kip
(D) 7 ft-kips/kip

128. A beam is simply supported over a 22 ft span and overhangs the left support 8 ft. Uniformly distributed dead loading of 2 kips/ft and live loading of 3 kips/ft are applied. The live load is positioned to produce maximum shear. The absolute maximum shear force at a point midway between the simple supports (point B′) is most nearly

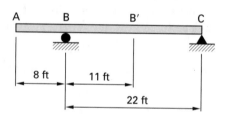

(A) 10 kips
(B) 16 kips
(C) 20 kips
(D) 39 kips

129. For the truss shown, the modulus of elasticity for all members is 29,000 ksi. The cross-sectional area of the members is 8 in². The horizontal deflection at joint D of the truss is most nearly

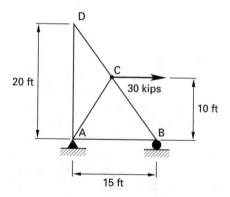

(A) 0.01 in
(B) 0.02 in
(C) 0.04 in
(D) 0.08 in

130. In the frame shown, the centroidal area moment of inertia for each leg is 650 in^4. The modulus of elasticity of steel reinforcement is 29,000 kips/in^3. Neglecting axial and shear deformation, the counterclockwise rotation at joint C is most nearly

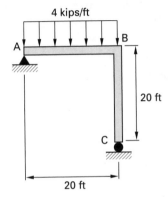

(A) 0.001 radians
(B) 0.004 radians
(C) 0.008 radians
(D) 0.01 radians

131. A pile group consists of 16 piles of the same size and type symmetrically arranged as shown. The group supports a vertical compressive force, P, of 800 kips at an eccentricity of 1.2 ft with respect to the centroid about one principal axis and 1.6 ft about the other principal axis. The maximum axial compression in a pile caused by this loading is most nearly

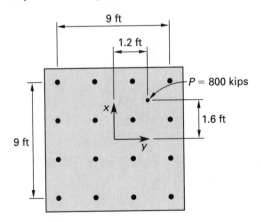

(A) 50 kips
(B) 60 kips
(C) 80 kips
(D) 90 kips

132. The plane truss shown is properly classified as

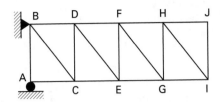

(A) statically determinate and stable
(B) statically determinate and unstable
(C) statically indeterminate to the 1st degree
(D) statically indeterminate to the 2nd degree

133. The two-span continuous beam shown is subject to a uniformly distributed load of 3.5 kips/ft over both spans. Support B experiences a differential settlement of 0.5 in downward relative to supports A and C. The beam has a moment of inertia of 1630 in^4 and modulus of elasticity of 29,000 kips/in^2 in both spans. The reaction at B caused by the uniform load and the differential settlement is most nearly

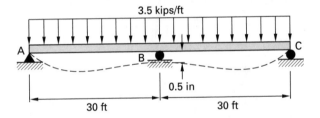

(A) 110 kips
(B) 120 kips
(C) 130 kips
(D) 140 kips

134. The three-span continuous steel beam shown is to be analyzed for the ultimate (i.e., factored) concentrated forces of 25 kips at the midpoint of each span. Neglecting member weight and assuming that strength is controlled by the formation of plastic hinges at critical locations, the required moment capacity is most nearly

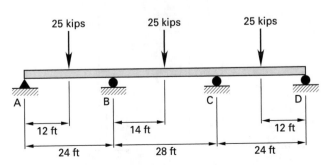

(A) 70 ft-kips
(B) 80 ft-kips
(C) 90 ft-kips
(D) 100 ft-kips

(A) 0.15 sec
(B) 0.3 sec
(C) 0.5 sec
(D) 0.8 sec

135. A 0.5 kip weight is dropped from rest from 4 ft above an initially straight, simply supported beam and contacts the beam at midspan. The beam is a steel section with a moment of inertia of 1600 in^4 about its axis of bending. Assuming linear elastic behavior, neglecting shear deformation, and considering the weight as an equivalent concentrated force at the midspan, the maximum force exerted on the beam is most nearly

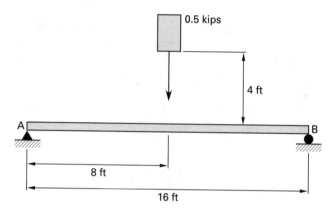

(A) 10 kips
(B) 20 kips
(C) 70 kips
(D) 120 kips

136. An elevated bulk storage bin weighs a total of 30 kips when filled to capacity. The bin can be idealized as a vertical cantilever supported by four steel columns that are fixed at the base and top. Each column has a moment of inertia about its axis of bending of 800 in^4 and is 16 ft long. The modulus of elasticity of steel is 29,000 kips/in^2. Assuming linear elastic behavior, the natural period of vibration for undamped lateral motion of the bin is most nearly

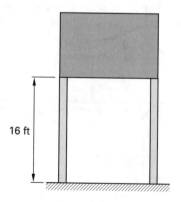

137. A normal weight prestressed concrete pile contains four $1/2$ in diameter, 270 ksi, concentrically placed strands that have a 3500 psi compressive stress at release of prestress. Strands are pretensioned to 200 ksi each immediately before being cut. The section is a solid 12 in × 12 in gross cross section. The loss of the prestress due to elastic shortening in this member is most nearly

(A) 5 kips/in^2
(B) 7 kips/in^2
(C) 8 kips/in^2
(D) 9 kips/in^2

138. A thin-walled steel pipe transports water at a pressure of 80 lbf/in^2. There is negligible restraint to the pipe in its longitudinal direction. If the pipe outside diameter is 60 in and the wall thickness is $3/8$ in, the change in pipe diameter caused by the internal pressure is most nearly

(A) 0.005 in
(B) 0.013 in
(C) 0.021 in
(D) 0.025 in

139. A steel column section is built by welding two plates 10 in wide by $5/8$ in thick to the flanges of a W12× 106 to form a doubly symmetrical section. The radius of gyration of the built-up section with respect to its major principal axis is most nearly

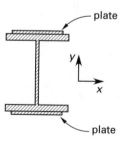

(A) 5.9 in
(B) 6.2 in
(C) 7.1 in
(D) 7.9 in

140. A W24 × 55 has coverplates 8 in wide by $1/2$ in thick symmetrically placed and welded to its top and bottom flanges. The section is subjected to a vertical shear force of 95 kips. Assuming linearly elastic behavior, the horizontal shear flow between the cover plate and flange is most nearly

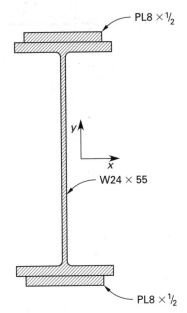

PL8 × ½

y

x

W24 × 55

PL8 × ½

(A) 2 kips/in
(B) 3 kips/in
(C) 5 kips/in
(D) 8 kips/in

141. A reinforced concrete retaining wall is backfilled with a granular material that exerts an active earth pressure equivalent to a 45 lbf/ft³ fluid. The backfill is level without a surcharge, and the stem extends vertically from the top of the backfill 13 ft to the top of the wall footing. The concrete is normal weight with specified compressive strength of 4000 psi, and grade 60 rebars are specified. The flexural steel percentage is specified to be 0.01. The required area of reinforcing steel at the bottom of the stem is most nearly

(A) 0.50 in²/ft
(B) 0.70 in²/ft
(C) 0.80 in²/ft
(D) 0.90 in²/ft

142. For normal weight concrete used in reinforced concrete structures that will be exposed to seawater, the maximum water-to-cementitious-materials ratio, by weight, is required to be

(A) 0.35
(B) 0.40
(C) 0.45
(D) 0.50

143. A reinforced concrete beam has a trough in the compression region as shown. Concrete is normal weight

with specified compressive strength of 4000 psi, and grade 60, no. 11 rebars are specified. Given that the steel yields when flexural failure occurs, the strain in the tension reinforcement when the compression strain in concrete reaches the ultimate value of 0.003 is most nearly

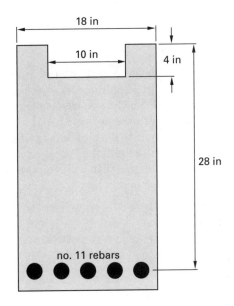

18 in

10 in

4 in

28 in

no. 11 rebars

(A) 0.004
(B) 0.005
(C) 0.006
(D) 0.007

144. A reinforced concrete column has three no. 9 rebars in each face as shown. The concrete is normal weight with specified compressive strength of 6000 psi, and rebars are grade 60. Under balanced strain conditions, the total compression force in the three rebars on the compression side is most nearly

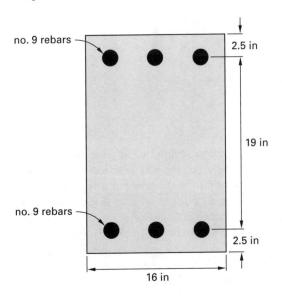

no. 9 rebars

2.5 in

19 in

no. 9 rebars

2.5 in

16 in

(A) 60 kips
(B) 90 kips
(C) 120 kips
(D) 180 kips

145. A reinforced concrete tied column is subjected to a design axial compression force of 1090 kips that is concentrically applied. Slenderness effects are negligible, and the column is to be designed using ACI 318. Given a specified compressive strength of 5000 psi, grade 60 rebars, and a specified longitudinal steel ratio of 0.02, the smallest square column that will support the load has sides that are most nearly how wide?

(A) 12 in
(B) 16 in
(C) 20 in
(D) 24 in

146. A reinforced concrete tied column is subjected to a design axial compression force of 875 kips and a design bending moment about its strong axis of 175 ft-kips. The column's cross section measures 20 in × 18 in, specified compressive strength is 4000 psi, steel is grade 60, and the distance from edge of column to center of steel in each face is 3 in. The required area of longitudinal steel is most nearly

(A) 4.0 in²
(B) 8.0 in²
(C) 12 in²
(D) 16 in²

147. A reinforced concrete corbel is to be designed to support a factored vertical reaction of 66 kips at an eccentricity of 6 in, measured from the face of the supporting column. The corbel is cast monolithically with the column, which is 16 in wide, and is of normal weight concrete with a specified compressive strength of 5000 psi and reinforced with grade 60 rebars. The required area of the primary steel reinforcement in the corbel, in accordance with ACI 318, is most nearly

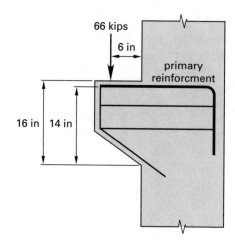

(A) 1.05 in²
(B) 1.25 in²
(C) 1.45 in²
(D) 1.65 in²

148. A combined footing constructed of normal weight reinforced concrete is subjected to concentrated forces from column reactions as shown. The columns are 12 in × 12 in, and the footing is 2 ft thick. The maximum bearing pressure beneath the footing, including the footing weight, is most nearly

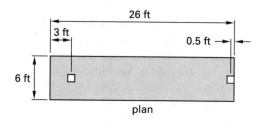

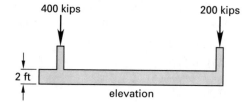

(A) 3 kips/ft²
(B) 4 kips/ft²
(C) 5 kips/ft²
(D) 6 kips/ft²

149. The cantilevered retaining wall shown retains soil with a unit weight of 100 lbf/ft³ that exerts active earth pressure equivalent to that of a fluid with unit weight 35 lbf/ft³. Concrete is normal weight. The factor of safety against overturning is most nearly

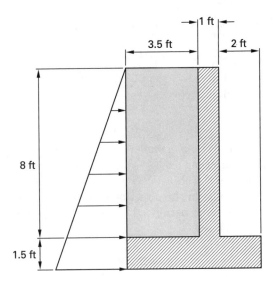

(A) 1
(B) 2
(C) 3
(D) 4

150. A rectangular beam is pretensioned using six $1/2$ in diameter grade 270 strands positioned as shown. The beam is nominally 40 ft long. The concrete is lightweight with a unit weight of 110 lbf/ft^3 and compressive strength of 3500 psi at time of release. The prestress is 200 kips/in^2 before release and is 180 kips/in^2 immediately after release. The area of prestressed reinforcement is $(6)(0.153 \text{ in}^2)$, or 0.918 in^2. The initial midspan upward camber in the beam is most nearly

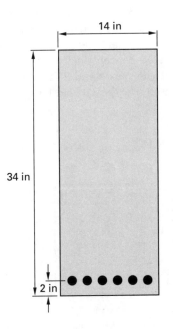

(A) 0.2 in
(B) 0.5 in
(C) 0.7 in
(D) 0.9 in

151. A two-span post-tensioned concrete beam has an effective prestress of 280 kips in the idealized profile shown. The rigidity of the member is 250×10^6 kips-in^2. The secondary downward reaction at support B induced by the prestress is most nearly

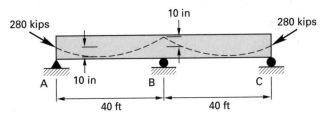

(not to scale)

(A) 3 kips
(B) 6 kips
(C) 9 kips
(D) 10 kips

152. The steel plate girder shown is compact against local buckling in A36 steel. The plastic moment capacity of the section for bending about its strong axis is most nearly

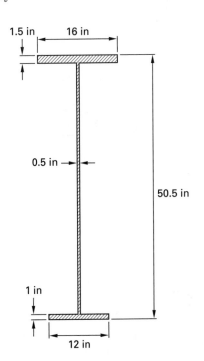

(A) 2200 ft-kips
(B) 2700 ft-kips
(C) 3300 ft-kips
(D) 3900 ft-kips

153. A steel W-section is uniformly loaded to produce bending about its strong axis. The beam supports a uniformly distributed service dead load of 2.5 kips/ft and a service live load of 1.8 kips/ft on a 36 ft simple span. Given that the beam is laterally supported only at the ends and at midspan, the maximum value of the bending coefficient that may be applied for this beam is most nearly

ASD option

(A) 1.2
(B) 1.4
(C) 1.6
(D) 1.8

LRFD option

(A) 1.0
(B) 1.1
(C) 1.2
(D) 1.3

154. Two L4 × 4 × ³/₈ are welded back to back to a gusset plate of A36 steel using ¹/₄ in E70 electrodes. The minimum thickness of the gusset plate that will develop the strength of the two welds is most nearly

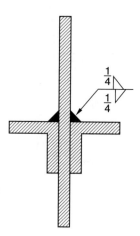

ASD option

(A) 0.3 in
(B) 0.4 in
(C) 0.5 in
(D) 0.6 in

LRFD option

(A) 0.2 in
(B) 0.3 in
(C) 0.4 in
(D) 0.6 in

155. A W10 × 49 member of A992 steel is punched for end connections using ³/₄ in diameter bolts through the flanges and web as shown. Assuming there are at least three bolts in the direction of stress, the allowable axial tensile force permitted by the AISC specification is most nearly

ASD option

(A) 350 kips
(B) 390 kips
(C) 430 kips
(D) 520 kips

LRFD option

(A) 510 kips
(B) 580 kips
(C) 650 kips
(D) 690 kips

156. A built-up compression member consists of two C12 × 30 symmetrically arranged and joined by two PL¹/₂ × 11, one on each side. The member is used as a column, loaded concentrically, with effective lengths of 22 ft about the x-axis and 24 ft about the y-axis. If A36 steel is used, the design load for the column is most nearly

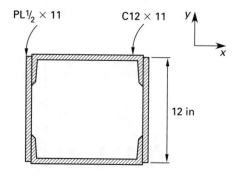

ASD option

(A) 550 kips
(B) 600 kips
(C) 650 kips
(D) 700 kips

LRFD option

(A) 800 kips
(B) 850 kips
(C) 900 kips
(D) 950 kips

157. A 24F glulam timber beam (with adjusted allowable bending stress of 2400 psi) spans 32 ft and is simply supported. The beam is braced laterally along its compression flange for half its length, but is not braced laterally over the other half, except at the support at the end of the beam. The beam is constructed using visually graded lumber and is $5^{1}/_{8}$ in wide by 22.5 in deep. It is subjected to loads of normal duration (i.e., 10 yr cumulative duration) and environmental conditions satisfying dry use. The elastic modulus with respect to the weak axis is 1.6×10^6 psi. The maximum bending moment that can be resisted by this member, based on the NDS, is most nearly

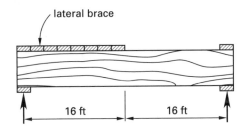

(A) 60 ft-kips
(B) 70 ft-kips
(C) 80 ft-kips
(D) 90 ft-kips

158. A diaphragm is constructed using Structural I grade plywood nailed to 2 in nominal wide framing in the pattern shown. The diaphragm must transfer lateral service wind of 320 lbf/ft to three shearwalls. The plywood thickness and nail pattern are adequate to transfer the maximum shear as a blocked diaphragm, and the plywood can transfer 240 lbf/ft as an unblocked diaphragm (with fasteners at their maximum spacing). The largest region between walls B and C (as measured from either side) over which an unblocked diaphragm is adequate is most nearly

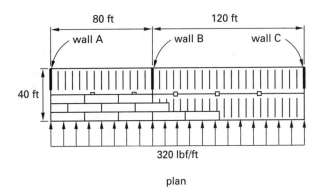

plan

(A) 10 ft to 110 ft
(B) 20 ft to 100 ft
(C) 30 ft to 90 ft
(D) 50 ft to 70 ft

159. A reinforced concrete masonry wall spans vertically 15 ft 4 in from the first floor to the roof. The wall is non-load-bearing and is governed by a wind pressure of 20 lbf/ft² acting inward or outward. The wall has a specified compressive strength of masonry of 1500 lbf/in² with special inspection, nominal 8 in masonry units grouted solid, grade 60 rebars, and bars centered in the wall. The required area of flexural reinforcement per foot of wall by IBC allowable stress design is most nearly

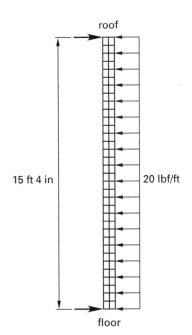

(A) 0.06 in²/ft
(B) 0.07 in²/ft
(C) 0.08 in²/ft
(D) 0.09 in²/ft

160. A W21 × 93 steel girder supporting a traveling crane is strengthened by welding $^3/_4$ in coverplates to top and bottom flanges using $^1/_4$ in E70 fillet welds. The welds are continuous and include transverse welds at the termination points, which occur 3 ft from each end of the girder. The girder is expected to experience approximately 50 applications of maximum stress daily during its design life of 25 yr. For these conditions, the stress range in the tension flange adjacent to the termination point is limited by the AISC specification to

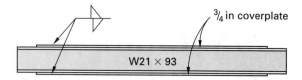

(A) 9 kips/in^2
(B) 12 kips/in^2
(C) 19 kips/in^2
(D) 21 kips/in^2

Afternoon Session
Transportation

161. The circular curve shown represents the centerline of a two-lane rural highway that passes through level surroundings. The highway has 12 ft lanes and 6 ft shoulders. The arc basis degree of curvature is 2°, and the design speed is 40 mph. The point of intersection (PI) is located at 423,968.68 N, 268,236.42 E. What are most nearly the coordinates of the point of tangent (PT)?

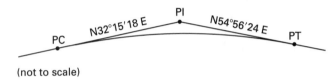

(not to scale)

(A) 424,297.26 N, 268,715.97 E
(B) 424,298.78 N, 268,706.80 E
(C) 424,309.47 N, 268,707.33 E
(D) 424,897.66 N, 268,136.44 E

162. Two tangents are connected by a circular curve as shown. The circular curve represents the centerline of a two-lane rural highway that passes through level surroundings. The highway has 12 ft lanes with 6 ft shoulders. The curve degree of curvature is 2°, and the design speed is 40 mph. What is most nearly the bearing of the radius line meeting the curve at PT?

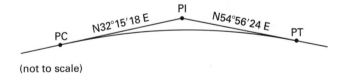

(not to scale)

(A) N 35°3'36" E
(B) N 54°5624" E
(C) S 34°2'16" E
(D) S 35°3'36" E

163. The centerline of a two-lane rural highway includes a circular curve as shown. The highway has 12 ft lanes with 6 ft shoulders. The curve degree of curvature is 2°, and the design speed is 40 mph. The curve is in the vicinity of an established Civil War cemetery. The point on the curve closest to the corner of the cemetery has coordinates 424,180.59 N and 268,549.70 E. What is most nearly the distance from the curve to the corner of the cemetery?

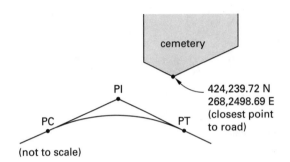

(not to scale)

(A) 77.5 ft
(B) 77.8 ft
(C) 78.1 ft
(D) 78.4 ft

164. A circular curve is designed as part of a two-lane rural highway that passes through level surroundings. The highway will have 12 ft lanes with 6 ft shoulders. The design speed is 40 mph and the curve degree of curvature is 2°. A reasonable recommendation of the resulting rate of superelevation is most nearly

(A) 0.02 ft/ft
(B) 0.03 ft/ft
(C) 0.06 ft/ft
(D) 0.08 ft/ft

165. A rural two-lane highway passes through level terrain. The highway has 12 ft lanes with 6 ft shoulders. The design speed is 40 mph. What is most nearly the approximate minimum required stopping sight distance?

(A) 92 ft
(B) 160 ft
(C) 310 ft
(D) 450 ft

166. The centerline of a circular curve in a two-lane roadway is shown. Each lane is 12 ft wide. There is no shoulder. The PC station is sta 12+40. The curve radius is 2080 ft. The interior angle is 60°. What is most nearly the PT station?

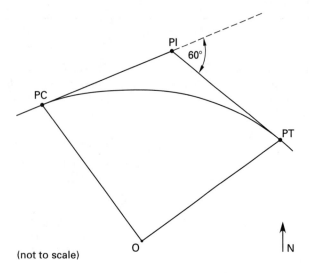

(not to scale)

(A) sta 24+38
(B) sta 22+18
(C) sta 34+18
(D) sta 48+18

167. A two-lane roadway is to be superelevated around a circular curve as shown. The axis of rotation will be the centerline. The criteria for the superelevation rate and runoff length are given in the following table. Tangent runouts are twice the runoff lengths listed. Each lane is 12 ft wide, and there is no shoulder. The PC station is sta 12+40. The curve radius is 2080 ft, and the interior angle is 60°. What is most nearly the station of the beginning of superelevation transition?

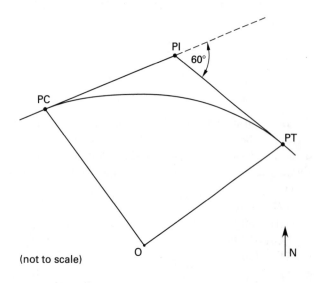

(not to scale)

curve radius	superelevation	runoff in the curve (ft)	
(ft)	(ft/ft)	2 lanes	4 lanes
22,920	NC	0	0
11,460	NC	0	0
7640	RC	0	0
5730	0.020	150	150
3820	0.028	150	150
2865	0.035	150	150
2290	0.040	150	150
1910	0.045	150	160
1640	0.048	150	170
1430	0.052	150	180
1145	0.056	150	200
955	0.059	150	210

(A) sta 6+40
(B) sta 9+40
(C) sta 11+40
(D) sta 11+90

168. A two-lane roadway includes a circular curve as shown. Each lane is 12 ft wide, and there is no shoulder. The curve length is 2000 ft, and the PC station is sta 13+50. The roadway is superelevated around the curve. The axis of rotation is at the centerline. The rate of superelevation is 0.045 ft/ft, and the runoff length in the curve is 150 ft. The roadway profile is on a constant uphill grade of 0.750%. The elevation of the centerline at the PC is 170 ft. What is most nearly the elevation of the outside (higher) edge of the roadway at the midpoint along the curve?

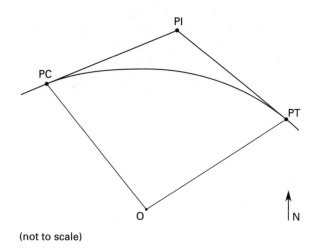

(not to scale)

(A) 170 ft
(B) 176 ft
(C) 177 ft
(D) 178 ft

169. The following traffic volume counts were taken on a road during the peak hour.

time period	volume (veh)
4:30–4:45 p.m.	195
4:46–5:00 p.m.	163
5:01–5:15 p.m.	157
5:16–5:30 p.m.	178

The peak hour factor (PHF) is most nearly

(A) 0.28
(B) 0.89
(C) 0.97
(D) 1.0

170. A six-lane rural freeway has an ideal free-flow speed of 70 mph. Each lane is 12 ft wide. A few sections have grades greater than 3%, but none of these sections are longer than $1/4$ mi. The directional weekday volume is 2500 vph in one direction and is mostly commuter traffic. The traffic stream consists of 14% trucks, 8% buses, and 4% recreational vehicles. The peak hour factor from previous volume studies is 0.85. The 15 min passenger car equivalent flow rate, under the prevailing conditions described, is most nearly

(A) 1100 pcphpl
(B) 1110 pcphpl
(C) 1170 pcphpl
(D) 1190 pcphpl

171. An eight-lane freeway near an urban area has an ideal 65 mph free-flow speed. There is an interchange approximately every 2 mi. Each lane is 11 ft wide. The minimum clear distance between overpass abutments, curve rails, and other roadside obstructions is 2 ft on both shoulders. A recent traffic study estimated the 15 min passenger car equivalent flow rate to be 1667 pcphpl. What is most nearly the level of service (LOS) at which the freeway is operating?

(A) LOS B
(B) LOS C
(C) LOS D
(D) LOS E

172. A new freeway with an ideal 70 mph free-flow speed is being designed. The freeway will pass through level terrain in suburban areas. The interchange frequency will be approximately one per mile. A traffic study estimates the 15 min passenger car equivalent flow rate to be 1874 pcphpl based on a six-lane freeway. What is most nearly the minimum number of lanes in each direction needed to provide level of service (LOS) C?

(A) 2
(B) 3
(C) 4
(D) 5

173. A six-lane freeway with an ideal 60 mph free-flow speed passes through rural areas. A traffic study estimates the current 15 min passenger car equivalent flow rate to be 1900 pcphpl. The flow rate is expected to grow at a rate of 5% per year. What is most nearly the number of years before the freeway starts operating at capacity?

(A) 1 yr
(B) 2 yr
(C) 3 yr
(D) 4 yr

174. The approach tangent to an equal-tangent vertical curve has a slope of +3%. The slope of the departure tangent is −2%. These two tangents intersect at sta 26+00 and elevation 231.00 ft. A set of subway rails passes below and perpendicular to the curve at sta 28+50 on the departure side. The maximum elevation of the railbed at that point is 195.00 ft. The length of the vertical curve is 16.48 sta. What is most nearly the minimum distance between the railbed and the roadway surface?

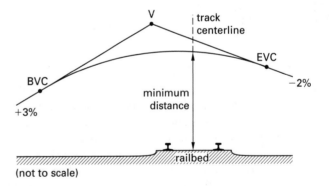

(not to scale)

(A) 5.0 ft
(B) 17 ft
(C) 26 ft
(D) 33 ft

175. A rural collector highway includes a crest vertical curve designed to connect a 2% grade with a −6% grade. The two vertical tangents intersect at sta 32+40.52 and elevation 456.61 ft. The design speed is 55 mph, and the actual traffic speeds are expected to be close to the design speed. What is most nearly the minimum length of the vertical curve for minimum stopping sight distance?

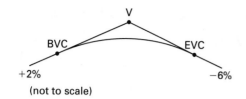

(not to scale)

(A) 300 ft
(B) 500 ft
(C) 700 ft
(D) 900 ft

176. A vertical curve connects two tangents. The approach tangent has a slope of +3%. The slope of the departure tangent is −2%. These two tangents intersect at sta 26+00 and elevation 231.00 ft. If the length of the vertical curve is 16.48 sta, what is most nearly the elevation of the EVC?

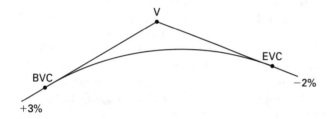

(A) 100 ft
(B) 200 ft
(C) 210 ft
(D) 250 ft

177. A four-lane highway runs north-south and passes through suburban areas. The preliminary design of the north-bound lanes includes a vertical curve with a length of 22.00 sta. The curve connects a +3% grade with a −5% grade. The two vertical tangents intersect at sta 91+70 and elevation 1453.61 ft. The design speed is 65 mph. What is most nearly the station of the highest point on the curve the turning point?

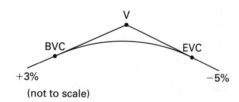

(A) sta 77+95
(B) sta 88+95
(C) sta 94+45
(D) sta 99+95

178. A two-lane highway is planned for a rural area. The horizontal layout of the highway includes four simple curves. The design criteria for the simple curves include a design speed of 60 mph and a rate of superelevation of 0.1. One of the simple curves shown is 1000 ft long with a radius of 1100 ft. The simple curve connects two tangents that deflect at an angle of 45° and intersect at sta 1500+00. What is most nearly the distance from the point of intersection of the two tangents (vertex) to the midpoint of the curve?

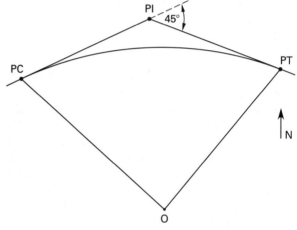

(not to scale)

(A) 82 ft
(B) 84 ft
(C) 91 ft
(D) 460 ft

179. The vertical alignment of a four-lane rural highway includes the vertical curve shown, which connects a −4% grade and a +1% grade. The point of vertical intersection of the grades occurs at sta 28+30 and has an elevation of 2231.31 ft. The design speed is 70 mph, and the length of the curve is 900 ft. What is most nearly the elevation of the lowest point on the curve (the turning point)?

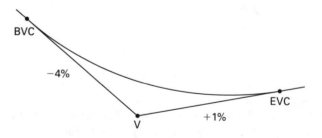

(A) 2221 ft
(B) 2231 ft
(C) 2233 ft
(D) 2235 ft

180. A 7 mi rural two-lane highway carries a two-way volume of 1100 vph. The highway passes through rolling terrain. The design includes 11 ft lanes, 4 ft paved shoulders, 40% no-passing zones, and 15 access points. A traffic study has shown that the directional split during peak hours is 60/40, and the peak hour factor is 0.92. The traffic consists of 5% trucks, 3% buses, and 2% recreational vehicles. The base free-flow speed is 60 mph. The highest directional flow rate for the peak 15 min period in pcph is most nearly

(A) 772 pcph
(B) 774 pcph
(C) 778 pcph
(D) 783 pcph

181. Plans outline a new four-lane freeway that will connect two cities through a suburban area. The freeway will serve as an alternative to an existing two-lane minor arterial, which is currently operating at capacity. The freeway will save motorists travel time and will be safer than the existing arterial. The demand function for travel on the new highway is represented by the straight line in the first illustration. The user cost per vehicle, including toll charges and delay costs, is represented by the supply curve in the second illustration. What is most nearly the expected number of vehicles to use the new freeway?

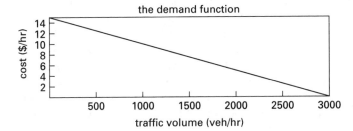

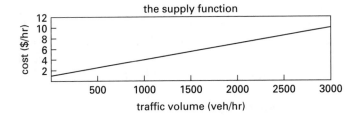

(A) 1000 vph
(B) 1250 vph
(C) 1500 vph
(D) 1750 vph

182. A highway under construction requires fill and cut based on the ground profile. The following mass diagram indicates the net accumulation of cut and fill between two stations. The maximum distance for which there are no additional hauling charges is 500 ft. The unit cost of excavation is $4.35/yd^3, and the overhaul unit cost is $9.75/yd^3 per station. What is most nearly the overhaul cost for the two stations shown?

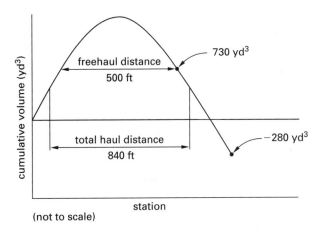

(A) $9300
(B) $11,000
(C) $24,000
(D) $60,000

183. Vehicles arrive at the ticket gate of a parking lot at an average rate of 30 vph. It takes an average of 1.5 min to get a ticket and drive through the gate. The arrival distribution is assumed to be Poisson and the service distribution is assumed to be exponential. There is no space limitation for the vehicles waiting to get a ticket. What is most nearly the number of vehicles expected to be waiting at the gate (i.e., the queue length)?

(A) one vehicle
(B) three vehicles
(C) four vehicles
(D) five vehicles

184. A traffic study area consists of four zones. The numbers of trip productions and attractions for each zone have been determined in the trip generation process as shown. A calibration process for the gravity model gives calibration values, $F_{i,j}$, which are a function of the travel time between each zone. The calibration values are shown. The socioeconomic conditions are considered to be the same in all zones. What is most nearly the number of trips that are produced in zone 1 and attracted to zone 3, using the gravity model for one iteration?

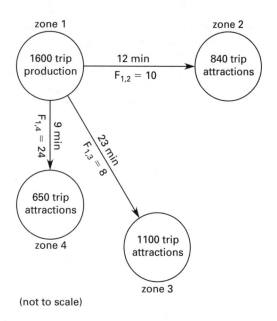

(not to scale)

(A) 300
(B) 430
(C) 740
(D) 1100

185. The following illustration shows a highway network consisting of nodes and links. The average travel times between links are shown. The following table gives the vehicle travel from node 1 to the other nodes based on an origin-destination (O-D) survey. All the links are two-way travel. Use the minimum travel time (all-or-nothing) approach to assign the traffic volumes generated in zone 1 to all available links. What is most nearly the amount of travel in vehicle-minutes for the link between nodes 1 and 5?

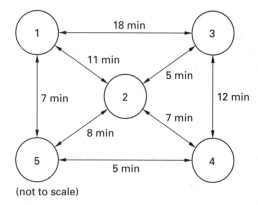

(not to scale)

	traffic volume (veh)
node 1 to node 1	0
node 1 to node 2	100
node 1 to node 3	85
node 1 to node 4	130
node 1 to node 5	115

(A) 800 veh-min
(B) 910 veh-min
(C) 1700 veh-min
(D) 4700 veh-min

186. The travel data in the following table relate to travel between two zones, zones 1 and 2, using either the auto mode or the transit mode in a suburban area. Use the trip interchange model (QRS method) to estimate the mode choice. The exponent value of the model is assumed to be 1.5, and the average income is $30,000.

	auto	transit
distance (mi)	12	10
cost per mile ($)	0.35	0.20
excess time (min)	6	11
speed (mph)	55	45

What is most nearly the expected percent of trips by auto, assuming 120,000 working minutes in a year?

(A) 43%
(B) 57%
(C) 64%
(D) 78%

187. A rural two-lane road has 9 ft lanes and 2 ft unpaved shoulders. A 15 mi section of this road has had 140 traffic accidents over the past 3 yr, 5% of which have been fatal accidents. The current average daily traffic (ADT) is 14,000 vpd. A nearby development is expected to increase the ADT on this section to 18,000 vpd. Construction associated with the development will improve the geometry of this section by adding 2 ft to the existing lanes and 2 ft to the existing shoulders. These improvements are expected to reduce the total and fatal traffic accidents by 25%. What is most nearly the fatal accident rate per 100 million veh-mi (HMVM) subsequent to the development and geometric improvements?

(A) 0.6 fatal accidents/HMVM
(B) 1.8 fatal accidents/HMVM
(C) 2.1 fatal accidents/HMVM
(D) 12 fatal accidents/HMVM

188. An intersection in a suburban area has had 54 traffic accidents in 1 yr, of which three were fatal and 13 resulted in injuries. The average 24 hr volumes entering each of the four approaches to the intersection are 1250 vpd, 2350 vpd, 730 vpd, and 1920 vpd. What is most nearly the injury accident rate per 10 million veh entering the intersection?

(A) 22 accidents/10^6 veh
(B) 57 accidents/10^6 veh
(C) 230 accidents/10^6 veh
(D) 240 accidents/10^6 veh

189. A parking garage is to be located in an urban business area. The garage will be open 6 a.m. to 5 p.m. A total of 400 vehicles will use the garage daily during the hours of operation. 75% of those who use the facility will be commuters with an average parking duration of 8 hr, and the rest are expected to be shoppers with an average parking duration of 3 hr. The parking efficiency, to account for turnovers, is expected to be 0.85. What will be most nearly the number of parking spaces required to meet the parking demand?

(A) 210 spaces
(B) 240 spaces
(C) 290 spaces
(D) 400 spaces

190. A curbed section of a two-lane road passes through an urban area. The section is 1.2 mi long and 40 ft wide curb to curb. Each lane is 11 ft wide. Parking is allowed on both sides of the road. The traffic consists mostly of passenger cars. An on-street parking configuration that minimizes interference with traffic movement is desired. What is most nearly the maximum number of parking spaces on this section of road?

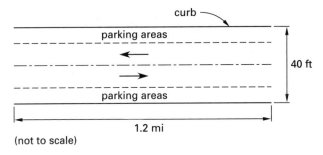

(not to scale)

(A) 290 spaces
(B) 580 spaces
(C) 700 spaces
(D) 740 spaces

191. A weaving section on a highway through an urban area serves the traffic flows shown. Lane widths are 12 ft with no lateral obstructions. The section is located in level terrain and has an average free-flow speed of 60 mph. The traffic flows represent ideal peak flows and are expressed in passenger cars per hour (pcph). The weaving section is assumed to operate in an unconstrained manner. What is most nearly the average speed of weaving vehicles?

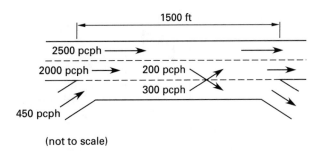

(not to scale)

(A) 38 mph
(B) 44 mph
(C) 51 mph
(D) 60 mph

192. An intersection located in an urban area has a four-phase signal. The four phases are shown. All phases are assumed to have the same lost time of 3 sec. For each phase, the ratios of actual flows to saturation flow, $(v/s)_{ci}$, for all critical lanes, groups, or approaches are also shown. The desired critical ratio of flow to capacity for the whole intersection is 0.90. What is most nearly the signal cycle length using *Highway Capacity Manual* (HCM) procedures?

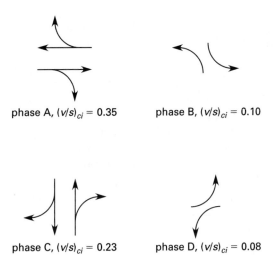

phase A, $(v/s)_{ci} = 0.35$ phase B, $(v/s)_{ci} = 0.10$

phase C, $(v/s)_{ci} = 0.23$ phase D, $(v/s)_{ci} = 0.08$

(A) 50 sec
(B) 78 sec
(C) 96 sec
(D) 120 sec

193. An intersection located in an urban area in level terrain is shown. The maximum allowable speed on the approach roads is 45 mph. The width of the intersection is 48 ft for all approaches. The average length of a vehicle is assumed to be 20 ft with a deceleration rate of 11.2 ft/sec^2. The perception-reaction time of drivers is 2 sec. What is most nearly the minimum yellow interval at this intersection?

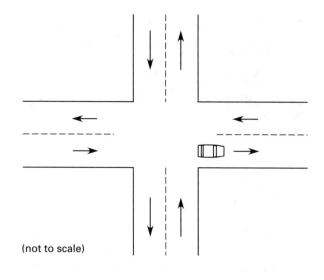

(not to scale)

(A) 2.5 sec
(B) 4.0 sec
(C) 5.5 sec
(D) 6.0 sec

194. Which one of the following statements is NOT true about the geometric design of horizontal curves?

(A) Reverse curves are not recommended because of the possibility of sudden changes to the alignment.
(B) Spiral curves are used between tangents and circular curves or between curves.
(C) Compound curves consist of two or more curves turning in opposite directions.
(D) The most common use of compound curves in highways is for at-grade intersections and ramps of interchanges.

195. Rapid-curing asphalt is formed by cutting asphalt cement with which one of the following distillates?

(A) heavy distillate such as diesel oil
(B) light fuel oil or kerosene
(C) petroleum distillate that easily evaporates, such as gasoline
(D) water

196. Which one of the following statements is NOT true in relation to soils for highway construction?

(A) The main reasons for soil compaction include minimizing future settlement and increasing soil strength.
(B) Sheepsfoot rollers are effective in compacting cohesive soils.
(C) A soil sample with an average CBR value of 48 is considered a superior subgrade to a soil with an average CBR value of 12.
(D) A soil classified A-7-6 (20) is usually rated "good" as a subgrade and is considered suitable as a subbase material.

197. The illustrations show results obtained using the Marshall mix method in designing an asphalt-concrete mixture. What is most nearly the optimum asphalt content for this mixture before checking design criteria for test limits?

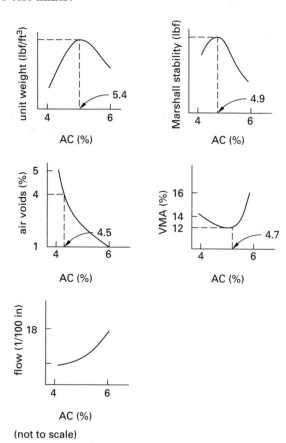

(not to scale)

(A) 4.5%
(B) 4.8%
(C) 5.0%
(D) 5.5%

198. A core is removed from a pavement containing an asphalt-concrete paving mixture designed with 6% asphalt cement. The mass of the core is 1238.5 g in air and 698.3 g submerged. The maximum specific gravity of the mixture is 2.48, and the bulk specific gravity of the combined aggregate in the mixture is 2.64. The percent voids in the mineral aggregate (VMA) is most nearly

(A) 7.7%
(B) 13%
(C) 18%
(D) 77%

199. A road will be constructed in a rural area where the traffic volume is expected to be low and the traffic speed fast. The road is located at latitude 40.9°. The 7 d average high air temperature is 31°C, and the 1 d average low air temperature is −25°C. The standard deviations of the high and low temperatures are ±1.5°C and ±2.7°C, respectively. The pavement surface depth will be 190 mm. Superpave^{TM} procedures will be used to design a suitable asphalt mixture. What will be an appropriate performance grade asphalt binder for this project for 98% reliability?

(A) PG 46-34
(B) PG 52-28
(C) PG 58-16
(D) PG 64-34

200. A flexible pavement for a major arterial road has been designed and applied. The following information is available.

material	thickness (in)	layer coefficient	drainage coefficient
AC surface course	6	0.400	–
untreated granular base	8	0.115	0.50
untreated gravel subbase	10	0.090	0.50

Among the pavement's other characteristics, the equivalent single axle load (ESAL) is 7×10^6; the standard deviation is 0.45; the subgrade resilient modulus is 1050 lbf/in^2; the CBR is 20 for the subbase and 50 for the base; and the elastic modulus of asphalt concrete is 350,000 lbf/in^2. The pavement structure is exposed to moisture levels approaching saturation for 16% of the time, and it takes approximately 70 days for drainage of water. The structural number (SN) of the constructed pavement is most nearly

(A) 2.8
(B) 3.3
(C) 4.2
(D) 4.5

Afternoon Session
Water Resources

---

201. The following table provides the test results for the discharge coefficients for sections of a broad-crested spillway.

ratio of design head to test head, H/H_o	ratio of design discharge coefficient to test discharge coefficient, C/C_o
0.2	0.84
0.4	0.90
0.6	0.93
0.8	0.97
1.0	1.00
1.2	1.04
1.4	1.04
1.6	1.09
1.8	1.11

The test discharge coefficient for a broad-crested design is 2.20 ft$^{1/2}$/sec for a head of 10 ft. The discharge for a design head of 16 ft and a length of 16 ft is most nearly

(A) 2000 ft^3/sec
(B) 2500 ft^3/sec
(C) 3000 ft^3/sec
(D) 3500 ft^3/sec

202. Which of the following statements are true with regard to energy dissipation below a dam?

I. If the upper conjugate depth of an hydraulic jump is below the tailwater, little energy will be dissipated.

II. An upturned bucket will protect the dam from scour by moving material toward the dam.

III. A secondary dam may increase tailwater height, thereby causing a hydraulic jump to form at the toe of the main dam.

IV. A sloping apron above streambed level may be used to control the hydraulic jump so that it occurs on the apron.

(A) I, II, III
(B) I, II, III, IV
(C) I, III, IV
(D) II, III

203. An outlet pipe from a reservoir (square edge inlet) consists of two sections of concrete pipe as shown. The discharge through a rotary valve (at point 3) is to the atmosphere. The water temperature is 50°F. What is most nearly the discharge through the valve?

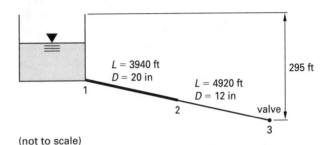

(not to scale)

(A) 4.0 ft^3/sec
(B) 9.0 ft^3/sec
(C) 13 ft^3/sec
(D) 18 ft^3/sec

204. The three-reservoir system described in the illustration and table is analyzed based on the flows at the junction of the pipes. Minor losses can be neglected, and the friction factor can be considered constant at all flows. Which of the following statements are correct?

I. The flow in pipe 1 flows into the junction.

II. The flow in pipe 1 flows out of the junction.

III. The flow in pipe 3 is 6.0 ft^3/sec.

IV. The flow in pipe 3 is 7.5 ft^3/sec.

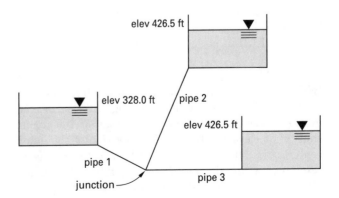

pipe	diameter (in)	length (ft)	friction factor, f	flow area (ft^2)
1	16	11,480	0.04	1.396
2	12	2950	0.04	0.785
3	20	5900	0.04	2.182

(A) I, III
(B) I, IV
(C) II, III
(D) II, IV

205. A newly developed 150 ac single-family residential subdivision is drained by an intermittent, straight stream with a uniform cross section. The area in flow is approximately rectangular, 4 ft wide, and 6 ft deep. The streambed is overgrown with weeds and has projecting stones. The elevation change is measured as 1.54 ft over a total length of $21 + 35.69$ sta. The capacity of the stream when flowing full is most nearly

(A) 36 ft^3/sec
(B) 59 ft^3/sec
(C) 73 ft^3/sec
(D) 120 ft^3/sec

206. Which of the following statements are correct?

I. Detention basins are characterized by ungated outlets.

II. Detention basins are usually designed to control short, high-intensity local storms.

III. Detention basins in the lower part of a river basin have little effect on reducing the flood crest from a storm moving downstream.

IV. Any number of detention basins will have little cumulative effect in reducing the peak discharge on the downstream section of a large river unless retention is provided over a prolonged period.

(A) I, II
(B) I, II, IV
(C) I, IV
(D) II, III

207. A pump station serving a small subdivision will lift water from a clear well of varying water surface elevation. The pump will start at elevation 100 ft and shut off at elevation 120 ft. The elevation of the pump centerline is 90 ft, and the elevation of the discharge reservoir is maintained at a constant 160 ft. The equivalent length of the discharge pipe is 2000 ft of 12 in diameter pipe, and the equivalent length of the suction pipe is 1500 ft of 16 in diameter pipe. Suction and discharge pipes are cement-lined cast iron. The pump performance data are given in the following table.

flow rate (gpm)	total dynamic head (ft)
500	96
1000	88
1500	76
2000	60
2500	36

The pump capacity range for the given conditions will be most nearly

(A) 600–1250 gpm
(B) 1100–1850 gpm
(C) 1350–1700 gpm
(D) 1650–2000 gpm

208. For the condition at which there is no flow into or out of the reservoir at node 2, which of the following statements are true for the pipe network system shown in the illustration?

I. The pressure head at node 6 will be the water surface elevation at the reservoir (100 ft) minus the pipe friction loss in pipes 5 and 6.

II. The flow in pipe 2 will always be $^2/_3 Q_1$.

III. The pressure head at node 2 will be 100 ft plus the pipe friction loss in pipe 1.

IV. The total head loss in pipes 1 and 2 must equal the total head loss in pipes 3 and 4.

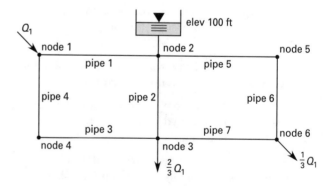

(A) I, II, IV
(B) I, III
(C) I, IV
(D) II, III, IV

209. A control structure on the outlet of a storm water retention pond contains a 1 ft diameter pipe set below a rectangular spillway as shown. The spillway has a coefficient of 0.62, and end contractions are suppressed. The pipe protrudes 2 ft into the retention pond. Assume a coefficient of discharge of 0.72. The water surface elevation, h, above the pipe centerline at which the spillway and the pipe will discharge at equal rates is most nearly

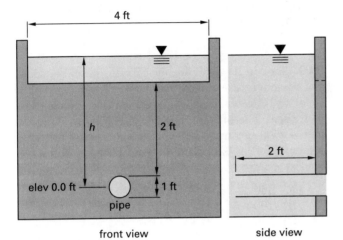

front view   side view

(not to scale)

(A) 1.5 ft
(B) 2.6 ft
(C) 3.2 ft
(D) 4.5 ft

210. A small community is being analyzed to determine its fire insurance rating. The domestic and commercial water demands and the fire rating demand characteristics of the community are given in the following list.

design population		9000	
average annual demand		180 gpcd	
large industrial demand		80 gpm in each zone	

	zone 1	zone 2	zone 3
average building separation, two-story single family	25 ft	40 ft	120 ft
essential service structures (hospital, police, emergency services)	20,000 ft^2 class 4	50,500 ft^2 class 5	40,800 ft^2 class 3

The minimum delivery capability that will best enable the community to meet domestic and industrial demands concurrently with the fire insurance rating standards for any zone is most nearly

(A) 2100 gpm
(B) 3100 gpm
(C) 4600 gpm
(D) 5100 gpm

211. A triangular culvert with an interior angle of 90° provides an open channel for conveyance of flood water under a highway. At a control structure 345 ft downstream, the water depth is 5.2 ft for a flow of 400 ft^3/sec. The channel is concrete lined and has a slope of 0.3%. Flow is not uniform, so it will be necessary to compute the length between 0.2 ft changes in water depth. The water depth at the highway crossing is most nearly

(A) 5.2 ft
(B) 5.5 ft
(C) 6.0 ft
(D) 6.4 ft

212. A pump supplies 10 ft^3/sec of water at 90°F from a pit with a constant water level at elevation 5000 ft. Suction friction loss is 3.6 ft at the operating point. The net positive suction head required (NPSHR) is 10.3 ft. What is most nearly the maximum height the pump centerline can be installed above the pit water level to provide a safety factor of 1.3 against cavitation?

(A) 6.0 ft
(B) 9.0 ft
(C) 12 ft
(D) 15 ft

213. A triangular channel with an interior angle of 90° carries a flow of 200 ft^3/sec. The actual depth of water in the channel is 3.2 ft. Which of the following statements are true?

   I. The flow is subcritical.

   II. The flow is supercritical.

   III. The water level is at the critical depth.

   IV. The Froude number is > 1.

   V. The Froude number is < 1

(A) I, IV
(B) I, V
(C) II, IV
(D) III, V

214. A rectangular open channel has the following characteristics and flow conditions.

width	9 ft
discharge	530 ft³/sec
channel slope	0.004
Manning coefficient	0.01

If a hydraulic jump takes place at the normal depth of flow, the energy loss through the jump is most nearly

(A) 0.09 ft
(B) 0.53 ft
(C) 0.96 ft
(D) 1.8 ft

215. A venturi meter is to be calibrated against a tank of known volume for which a discharge is timed. The tank volume is 6.66 ft³ and is totally emptied in 2 sec. The venturi meter characteristics are as follows.

throat diameter	6 in
manometer fluid	mercury
height of manometer fluid	8 in
temperature of water	50°F

If the peak discharge is 150% of the average, the flow coefficient of the meter is most nearly

(A) 0.76
(B) 0.84
(C) 0.91
(D) 0.98

216. A 2.5 m diameter concrete pipe is being analyzed to determine its maximum discharge capacity. The head loss has been determined to be 1.5 m per kilometer of pipe length at a temperature of 10°C when using the best condition of the pipe. The maximum discharge capacity of the pipe is most nearly

(A) 10 m³/s
(B) 20 m³/s
(C) 30 m³/s
(D) 40 m³/s

217. A culvert for a highway fill is to carry a peak discharge of 30 ft³/sec, at optimum discharge characteristics, through a 24 in, very smooth concrete pipe with a Manning roughness constant of 0.012. The minimum slope that will achieve this condition is most nearly

(A) 0.008 ft/ft
(B) 0.013 ft/ft
(C) 0.018 ft/ft
(D) 0.020 ft/ft

218. Flow under a vertical gate (gate discharge coefficients $\Psi = 0.624$ and $C = 0.603$) is controlled by adjusting the gate opening. For an upstream water depth of 6 ft and velocity of 10 ft/sec in a 2 ft wide rectangular channel with a 2 ft opening, the downstream free flow velocity is most nearly

(A) 5.0 ft/sec
(B) 10 ft/sec
(C) 15 ft/sec
(D) 20 ft/sec

219. An 8 in ID steel pipeline carrying gasoline, which has a temperature of 60°F and a specific gravity of 0.80 (relative to water at 60°F), experiences a valve closure in 1.2 sec. The pipeline has a wall thickness of 0.25 in and is 2500 ft long. The valve closure occurs at a flow velocity of 10 ft/sec. The pressure rise that will occur is most nearly

(A) 140 psi
(B) 190 psi
(C) 260 psi
(D) 310 psi

220. Water is released from a wide sluice gate into a rectangular stilling basin of the same width. The bottom of the basin is horizontal and is constructed of concrete. The average discharge velocity is 62 ft/sec, measured a short distance downstream from the point of minimum discharge depth. The minimum depth is 4.08 ft. Beyond the measurement point, the depth increases gradually. Somewhere downstream, the flow experiences a hydraulic jump to attain the depth of the stilling basin. The depth of the flow just before the hydraulic jump is 4.57 ft. The approximate depth of the stilling basin after the hydraulic jump is most nearly

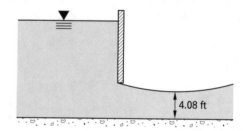

(A) 6.3 ft
(B) 7.9 ft
(C) 25 ft
(D) 27 ft

221. The annual precipitation at station C is being used to characterize a storm at this station. The station was moved in 1960. The annual precipitation at station C and the precipitation at a group of 10 surrounding stations are given in the following table.

year	station C annual precipitation (in)	adjacent 10 stations' mean of annual precipitation (in)
1950	–	–
1951	25.0	30.0
1952	24.6	31.2
1953	26.3	29.7
1954	25.7	30.5
1955	24.3	28.7
1956	25.9	31.3
1957	27.3	30.6
1958	24.5	29.9
1959	23.9	29.8
1960	22.5	28.3
1961	30.0	20.0
1962	31.2	21.2
1963	29.8	19.9
1964	30.5	20.5
1965	28.5	18.4

Which one of the following statements is true?

(A) Station C data is consistent with the group of stations since the rate of rainfall accumulations is constant for both from 1950 to 1965.

(B) Station C is not consistent with the 10 stations due to a change in the rate of rainfall accumulations.

(C) Not enough information is given to determine if station C data is consistent with the 10 stations data.

(D) Station C data is consistent with the 10 station data, even though the rate of change of rainfall accumulation varies for both.

222. The precipitation at four rain gauge stations in a small watershed is given in the following table.

station	annual precipitation (in)	Storm Alpha precipitation (in)	duration (min)	contributing watershed area (ac)
A	40.5	3.20	57	2000
B	32.8	2.70	57	2500
C	60.7	3.90	57	2800
D	50.3	?	57	2400

The return frequency of storms for this watershed can be found from the following intensity-duration-frequency curves. The return frequency of Storm Alpha is most nearly

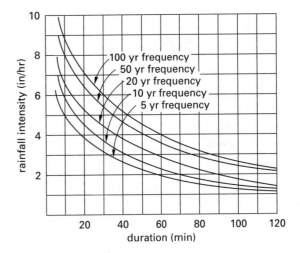

(A) 5 yr
(B) 10 yr
(C) 20 yr
(D) 50 yr

223. Which of the following statements are true?

I. Unit hydrographs are based on the assumption that only the total amount of rainfall varies from storm to storm.

II. Construction of a unit hydrograph requires an overland flow hydrograph and known values for the area and time of concentration of the drainage basin.

III. A synthetic hydrograph can be constructed for ungauged drainages using the U.S. Natural Conservation Service curve number method.

IV. If a storm's duration is not the same as or close to the hydrograph base length, the unit hydrograph cannot be used to predict runoff.

V. Methods to develop a unit hydrograph for durations of storms that are multiples of the storm duration for which a unit hydrograph is available include the rational method, the lagging storm method, and the S-curve method.

(A) I, II, IV
(B) I, III, IV
(C) II, III, V
(D) III, IV, V

224. A small watershed is covered predominately in alfalfa during the month of May. The watershed has a crop coefficient, k_c, of 0.35 and a soil coefficient, k_s, of 0.80 for a soil with good watering characteristics. The mean wind run in May is 260 km/d, and the mean temperature is 20°C. The mean relative humidity for May is 30% for the watershed. The net radiation for May is 250 W/m². The weighting factors for radiation and vapor transport are 0.7 and 0.3, respectively. The evapotranspiration for the alfalfa watershed is most nearly

(A) 1 mm/d
(B) 3 mm/d
(C) 6 mm/d
(D) 8 mm/d

225. A soil permeation test determined a saturated hydraulic conductivity rate of 1 in/hr for a test apparatus of 6 in² of exposed soil. A total of 12 in³ of water was used in the soil suction test, which was conducted for 20 min. The infiltration depth that would occur for the soil after 1 hr would be most nearly

(A) 2.0 in
(B) 4.0 in
(C) 8.0 in
(D) 12 in

226. Maximum total precipitation data for a small watershed are given in the following table for 5 min and 20 min subdurations and 10 yr of record.

5 min subduration (cm)	20 min subduration (cm)
1.8	2.1
1.2	2.4
1.0	2.3
1.9	3.1
0.8	2.9
0.7	2.5
1.3	2.8
1.5	2.6
1.2	3.2
0.9	2.4

For a 15 min duration storm, the percent increase in intensity from a 2-yr to a 10-yr storm is most nearly

(A) 10%
(B) 20%
(C) 30%
(D) 40%

227. Sheet flow occurs across a large, impervious parking surface that is smooth and uniform. The surface is 150 ft along the direction of a 6% slope. For a temperature of 60°F and a rainfall intensity of 1.3 in/hr, the velocity of flow at the end of the parking surface is most nearly

(A) 0.20 ft/sec
(B) 0.80 ft/sec
(C) 1.3 ft/sec
(D) 2.2 ft/sec

228. Precipitation data are unavailable at a proposed development area in a watershed. However, precipitation data at four stations surrounding the development are complete and reliable. The available data are given in the following table.

station	10-yr, 30 min intensity (in/hr)	distance to development (mi)
A	3.2	6.3
B	4.6	4.2
C	2.9	5.1
D	3.7	8.1

The estimated 10-yr, 30 min intensity at the development would be most nearly

(A) 3.1 in/hr
(B) 3.5 in/hr
(C) 3.8 in/hr
(D) 4.0 in/hr

229. A natural channel and flood plain are shown in the following illustration. During a flood, the river overflows its banks into the east and west portions of the natural flood plain. The river channel is in good condition, but the west flood plain is filled with stones and weeds, and the east flood plain is very poor. The channel and flood plain slopes are approximately 0.002 ft/ft. During a steady flow period when the water depth can be measured as shown, the total flood discharge is most nearly

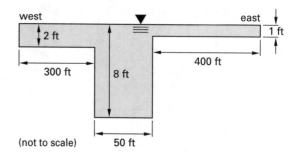

(not to scale)

(A) 4000 ft^3/sec
(B) 5000 ft^3/sec
(C) 6000 ft^3/sec
(D) 7000 ft^3/sec

230. Untreated water is stored in an uncovered municipal reservoir with an average winter depth of 15 ft and an average winter temperature of 40°F. During the summer, the land surrounding the reservoir accumulates wind-blown silt (approximately spherical particles with a specific gravity of 2.65) and organic debris. The first winter storm washes all of this material into the reservoir. After the storm, the floating organic debris is quickly skimmed off, leaving the silt in the water to settle out. The water is considered clear enough to withdraw when the largest particle in suspension has a size of 4.0×10^{-4} in. There is adequate storage elsewhere to provide treated water for 10 hr after a storm. The maximum time needed for the reservoir to become clear enough to be withdrawn is most nearly

(A) 6 hr
(B) 8 hr
(C) 10 hr
(D) 20 hr

231. The future growth of the community of Bountiful is limited by economic and geographic factors to a limited saturated population as represented by an S-shaped growth curve. Analysis indicates the current population of 100,000 will grow as represented by a k constant of -0.0055 and a saturated population of 300,000 for future growth. The per capita water demand is currently 150 gal/capital-day and is expected to increase at 10% of the percent population increase. The total water demand in 40 yr will be most nearly

(A) 20 MGD
(B) 35 MGD
(C) 45 MGD
(D) 60 MGD

232. A water treatment plant with the unit processes and design criteria given in the following table is being analyzed for deficiencies.

unit process	design criteria	rated capacity
screening	1.25 maximum hourly demand	300 000 m^3/d
influent flow meter	1.5 maximum hourly demand	14 400 m^3/h
flash mix	1.25 maximum daily demand	20 MGD
flocculation basin	1.25 maximum daily demand	25,000 gpm
sedimentation basin	1.5 maximum daily demand	2×10^6 gal/hr
filters	1.0 maximum daily demand	50 000 L/min
effluent flow meter	1.5 maximum hourly demand	10 000 m^3/h
chlorine contact/ clearwell	1.5 maximum daily demand	70 ft^3/sec
high service pumps	1.0 maximum hourly demand	30,000 gpm

The maximum daily demand is 150% of the average daily demand, and the maximum hourly demand is 200% of the maximum daily demand. The annual average daily design demand is 800 L/s. The unit processes that require upgrading to meet the design criteria are

(A) screening, flash mix, filters, chlorine contact/ clearwell
(B) flash mix, filters, chlorine contact/clearwell, high service pumps
(C) screening, influent flow meter, flocculation basin, sedimentation basin
(D) flash mix, filters, effluent flow meter, high service pumps

233. The size of a raw water storage reservoir is designed to meet a 2 yr drought condition with negligible precipitation. The stream flow available for use by the treatment plant was analyzed for the low flow conditions as shown in the following table for selected subdurations. Also shown are the evaporation/seepage losses and the design demands for the corresponding subdurations. The area of the reservoir after construction will be 4 km^2.

subduration (d)	average inflow rate (Mm^3/d)	evaporation/ seepage rate (mm)	design demand rate (Mm^3/d)
7	0.120	30	0.995
30	0.150	120	0.990
60	0.180	250	0.980
120	0.240	500	0.960
180	0.560	780	0.940
365	1.800	1600	0.900

The required volume of the raw water storage reservoir is most nearly

(A) 22 Mm³
(B) 53 Mm³
(C) 76 Mm³
(D) 88 Mm³

234. Which of the following are true with respect to rapid (flash) mixing?

I. Dispersion of metal coagulants should be completed in a fraction of a second due to hydrolysis of the coagulant.

II. Flash mixing of anionic polymers can be completed in 2–3 sec because hydrolytic reactions are not involved.

III. The pH of the process water is seldom critical or important in flash mixing of coagulants.

IV. The sequence of chemical addition is important in rapid-mixing operations.

V. In-line static mixers offer the advantages of effective mixing and constant mixing time.

(A) I, IV
(B) I, IV, V
(C) I, V
(D) II, III, IV

235. A water treatment plant uses alum for coagulation-flocculation at a dose of 12 mg/L. The stoichiometric mass of alum sludge produced is 0.46 kg of sludge per kg of alum dose. The raw water has a turbidity of 5 NTU, and the turbidity after sedimentation is 1 NTU. Jar tests and other studies determined this water has a correlation between turbidity and suspended solids removal, which is not usually the case. The total suspended solids removed is a function of the turbidity removed according to the following formula.

$$\Delta TSS \text{ in mg/L} = \Delta NTU^{1.2}$$

As a coagulant aid, 3 mg/L of clay are added. The removal efficiency of the sedimentation unit is 96% for this range of turbidity. For a flow of 2 m³/s, the mass of solids removed is most nearly

(A) 1500 kg/d
(B) 2300 kg/d
(C) 3200 kg/d
(D) 4300 kg/d

236. A clarifier for flocculated sedimentation is being designed for a new water treatment plant using a coagulant X on the raw water. Batch settling tests were performed in a 10 ft column. A graph of suspended solids removal (as percent) versus depth and settling times is shown. For a design flow of 1.5 MGD and a scale-up factor of 1.0, what are most nearly the diameter and depth, respectively, of a clarifier that can provide 75% removal?

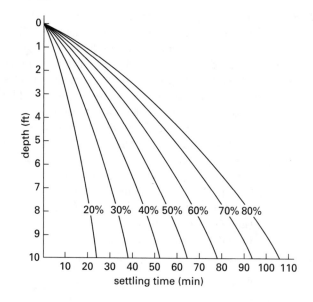

(A) 32 ft, 6 ft
(B) 37 ft, 10 ft
(C) 42 ft, 8 ft
(D) 45 ft, 12 ft

237. A dual-media filter has the characteristics given in the following table. Assume each layer is uniform.

parameter	property	
	sand	garnet
depth	280 mm	100 mm
SG	2.65	4.20
average size	1.0 mm	0.40 mm
porosity, ξ	0.50	0.55
shape factor, ϕ	0.8	0.8

For a filtration rate of 4.5 L/s·m² and a temperature of 10°C, the head loss through a clean filter is most nearly

(A) 0.12 m
(B) 0.18 m
(C) 0.26 m
(D) 0.34 m

238. A clearwell is to be disinfected before being placed in service. The clearwell is 1000 m³ in volume. A 0.5%

hypochlorite solution is to be used. The mass of 70% available dry hypochlorite powder required is most nearly

(A) 4000 kg
(B) 5500 kg
(C) 6500 kg
(D) 7000 kg

239. Which of the following statements are true?

I. Finalized USEPA rules applicable to water treatment design include the Surface Water Treatment Rule; the Total Coliform Rule; the Phase I, II, and V Contaminant rules; and the Lead and Copper Rule.

II. Primary drinking water standards can only be enforced by USEPA.

III. Secondary drinking water standards can be enforced by a state drinking water regulatory agency if the agency has appropriate state legal authority.

IV. The maximum contaminant level for turbidity is 1.0 NTU in a maximum of 5% of monthly samples.

V. The maximum four-quarter average primary standard for trihalomethanes is 0.10 mg/L.

VI. Microbiological contamination shall not exceed one positive sample per month.

(A) I, II, VI
(B) I, II, V
(C) I, III, V
(D) II, III, IV

240. A 2 yr membrane test to produce drinking water from a high solids, brackish feed water yielded the data in the following table.

time (hr)	flux (gal/day-ft^2)
10	24.2
10,000	7.0
20,000	6.2

The percent decrease in performance between 2 yr and 3 yr of operation would be most nearly

(A) 5%
(B) 7%
(C) 9%
(D) 11%

STOP!

DO NOT CONTINUE!

This concludes the Afternoon Session of the examination. If you finish early, check your work and make sure that you have followed all instructions. After checking your answers, you may turn in your examination booklet and answer sheet and leave the examination room. Once you leave, you will not be permitted to return to work or change your answers.

Answer Key

Morning Session

1. A B C **D**	11. A B C **D**	21. A **B** C D	31. A B **C** D
2. A **B** C D	12. A **B** C D	22. A B **C** D	32. A **B** C D
3. A **B** C D	13. A B **C** D	23. A **B** C D	33. A B C **D**
4. A B **C** D	14. A B C **D**	24. A B **C** D	34. A B **C** D
5. A B **C** D	15. A **B** C D	25. A **B** C D	35. A **B** C D
6. A **B** C D	16. **A** B C D	26. A B **C** D	36. A B C **D**
7. A **B** C D	17. **A** B C D	27. A **B** C D	37. A **B** C D
8. **A** B C D	18. A B **C** D	28. A **B** C D	38. **A** B C D
9. A **B** C D	19. **A** B C D	29. A B **C** D	39. A B C **D**
10. A B **C** D	20. A **B** C D	30. A B **C** D	40. A **B** C D

Afternoon Session—Environmental

41. C	51. B	61. B	71. A
42. D	52. D	62. B	72. B
43. C	53. D	63. D	73. A
44. C	54. D	64. B	74. C
45. B	55. B	65. D	75. C
46. D	56. A	66. C	76. D
47. A	57. D	67. B	77. D
48. D	58. D	68. C	78. B
49. C	59. C	69. B	79. A
50. D	60. B	70. D	80. C

Afternoon Session—Geotechnical

81. B	91. D	101. D	111. B
82. A	92. D	102. C	112. A
83. B	93. C	103. D	113. B
84. B	94. A	104. D	114. A
85. D	95. B	105. A	115. B
86. C	96. C	106. C	116. C
87. C	97. D	107. C	117. A
88. B	98. D	108. B	118. D
89. B	99. C	109. C	119. C
90. C	100. D	110. A	120. B

Afternoon Session—Structural

121. C	131. C	141. D	151. B
122. C	132. B	142. B	152. C
123. B	133. C	143. A	153. D
124. B	134. D	144. D	154. C
125. D	135. D	145. C	155. B
126. C	136. A	146. B	156. A
127. C	137. B	147. A	157. A
128. B	138. B	148. D	158. C
129. C	139. A	149. D	159. B
130. D	140. A	150. B	160. A

Afternoon Session—Transportation

161. Ⓐ ● Ⓒ Ⓓ	171. Ⓐ Ⓑ ● Ⓓ	181. Ⓐ Ⓑ Ⓒ ●	191. Ⓐ ● Ⓒ Ⓓ
162. Ⓐ Ⓑ Ⓒ ●	172. Ⓐ Ⓑ ● Ⓓ	182. Ⓐ Ⓑ ● Ⓓ	192. Ⓐ ● Ⓒ Ⓓ
163. Ⓐ Ⓑ ● Ⓓ	173. Ⓐ Ⓑ Ⓒ ●	183. Ⓐ ● Ⓒ Ⓓ	193. Ⓐ Ⓑ Ⓒ ●
164. Ⓐ ● Ⓒ Ⓓ	174. Ⓐ Ⓑ ● Ⓓ	184. Ⓐ ● Ⓒ Ⓓ	194. Ⓐ Ⓑ ● Ⓓ
165. Ⓐ Ⓑ ● Ⓓ	175. Ⓐ Ⓑ Ⓒ ●	185. Ⓐ Ⓑ ● Ⓓ	195. Ⓐ Ⓑ ● Ⓓ
166. Ⓐ Ⓑ ● Ⓓ	176. Ⓐ Ⓑ ● Ⓓ	186. Ⓐ ● Ⓒ Ⓓ	196. Ⓐ Ⓑ Ⓒ ●
167. ● Ⓑ Ⓒ Ⓓ	177. Ⓐ ● Ⓒ Ⓓ	187. Ⓐ ● Ⓒ Ⓓ	197. Ⓐ Ⓑ ● Ⓓ
168. Ⓐ Ⓑ Ⓒ ●	178. Ⓐ Ⓑ ● Ⓓ	188. Ⓐ ● Ⓒ Ⓓ	198. Ⓐ Ⓑ ● Ⓓ
169. Ⓐ ● Ⓒ Ⓓ	179. Ⓐ Ⓑ Ⓒ ●	189. Ⓐ Ⓑ ● Ⓓ	199. Ⓐ Ⓑ ● Ⓓ
170. Ⓐ ● Ⓒ Ⓓ	180. Ⓐ Ⓑ ● Ⓓ	190. Ⓐ ● Ⓒ Ⓓ	200. Ⓐ ● Ⓒ Ⓓ

Afternoon Session—Water Resources

201. Ⓐ ● Ⓒ Ⓓ	211. Ⓐ Ⓑ ● Ⓓ	221. Ⓐ ● Ⓒ Ⓓ	231. ● Ⓑ Ⓒ Ⓓ
202. Ⓐ ● Ⓒ Ⓓ	212. Ⓐ ● Ⓒ Ⓓ	222. Ⓐ Ⓑ Ⓒ ●	232. Ⓐ Ⓑ Ⓒ ●
203. Ⓐ ● Ⓒ Ⓓ	213. Ⓐ Ⓑ ● Ⓓ	223. Ⓐ ● Ⓒ Ⓓ	233. Ⓐ Ⓑ Ⓒ ●
204. Ⓐ ● Ⓒ Ⓓ	214. ● Ⓑ Ⓒ Ⓓ	224. Ⓐ ● Ⓒ Ⓓ	234. ● Ⓑ Ⓒ Ⓓ
205. ● Ⓑ Ⓒ Ⓓ	215. Ⓐ Ⓑ ● Ⓓ	225. Ⓐ ● Ⓒ Ⓓ	235. Ⓐ ● Ⓒ Ⓓ
206. Ⓐ ● Ⓒ Ⓓ	216. ● Ⓑ Ⓒ Ⓓ	226. Ⓐ Ⓑ Ⓒ ●	236. Ⓐ ● Ⓒ Ⓓ
207. Ⓐ Ⓑ Ⓒ ●	217. Ⓐ ● Ⓒ Ⓓ	227. Ⓐ ● Ⓒ Ⓓ	237. Ⓐ Ⓑ Ⓒ ●
208. Ⓐ Ⓑ ● Ⓓ	218. Ⓐ Ⓑ Ⓒ ●	228. Ⓐ ● Ⓒ Ⓓ	238. Ⓐ Ⓑ Ⓒ ●
209. Ⓐ Ⓑ ● Ⓓ	219. Ⓐ Ⓑ Ⓒ ●	229. Ⓐ Ⓑ ● Ⓓ	239. Ⓐ Ⓑ ● Ⓓ
210. Ⓐ Ⓑ Ⓒ ●	220. Ⓐ Ⓑ Ⓒ ●	230. Ⓐ Ⓑ Ⓒ ●	240. Ⓐ ● Ⓒ Ⓓ

Solutions
Morning Session

1. The mixed liquor volatile suspended solids (MLVSS) is

$$\overline{X} = PC$$

C is the concentration of mixed liquor suspended solids (MLSS). P is the percent volatile solids.

$$\overline{X} = \left(0.75 \, \frac{\text{MLVSS}}{\text{MLSS}}\right)\left(2500 \, \frac{\text{mg}}{\text{L}} \, \text{MLSS}\right)$$
$$= 1875 \text{ mg/L MLVSS}$$

The influent biodegradable COD is

$$S_i = \text{influent COD} - \text{nonbiodegradable COD}$$
$$= 1800 \, \frac{\text{mg}}{\text{L}} - 110 \, \frac{\text{mg}}{\text{L}}$$
$$= 1690 \text{ mg/L}$$

The effluent biodegradable COD is

$$S_t = \text{effluent COD} - \text{nonbiodegradable COD}$$
$$= 200 \, \frac{\text{mg}}{\text{L}} - 110 \, \frac{\text{mg}}{\text{L}}$$
$$= 90 \text{ mg/L}$$

The reaction time is

$$\theta = \frac{S_i - S_t}{K\overline{X}S_t}$$
$$= \frac{1690 \, \frac{\text{mg}}{\text{L}} - 90 \, \frac{\text{mg}}{\text{L}}}{\left(0.6 \, \frac{\text{L}}{\text{g·h}}\right)\left(1875 \, \frac{\text{mg}}{\text{L}}\right)\left(\frac{1 \text{ g}}{1000 \text{ mg}}\right)\left(90 \, \frac{\text{mg}}{\text{L}}\right)}$$
$$= 15.8 \text{ h}$$

The reactor volume is

$$V = Q\theta$$
$$= \left(8 \times 10^6 \, \frac{\text{gal}}{\text{day}}\right)\left(\frac{1 \text{ ft}^3}{7.48 \text{ gal}}\right)(15.8 \text{ hr})\left(\frac{1 \text{ day}}{24 \text{ hr}}\right)$$
$$= 704{,}100 \text{ ft}^3 \quad (700{,}000 \text{ ft}^3)$$

The answer is (D).

2. The fraction of suspended solids removed is

$$\eta = \frac{C_i - C_o}{C_i} = \frac{500 \, \frac{\text{mg}}{\text{L}} - 150 \, \frac{\text{mg}}{\text{L}}}{500 \, \frac{\text{mg}}{\text{L}}}$$
$$= 0.70$$

The flow to each unit is

$$Q_{\text{unit}} = \frac{Q_{\text{total}}}{\text{no. units}} = \frac{15 \text{ MGD}}{2 \text{ units}}$$
$$= 7.5 \text{ MGD} \quad (7.5 \times 10^6 \text{ gal/day})$$

Using the pilot plant graphed results for 70% suspended solids removed, the overflow rate, v^*, is 1900 gal/day-ft^2, and the detention time is approximately 33 min. The diameter of each unit is

$$D = \sqrt{\frac{4Q}{\pi v^*}}$$
$$= \sqrt{\frac{(4)(7.5 \text{ MGD})\left(10^6 \, \frac{\text{gal}}{\text{MG}}\right)}{\pi\left(1900 \, \frac{\text{gal}}{\text{day-ft}^2}\right)}}$$
$$= 70.9 \text{ ft} \quad (70 \text{ ft})$$

The tank volume is

$$V = t_d Q$$
$$= (33 \text{ min})(7.5 \text{ MGD})\left(10^6 \, \frac{\text{gal}}{\text{MG}}\right)\left(\frac{1 \text{ ft}^3}{7.48 \text{ gal}}\right)$$
$$\times \left(\frac{1 \text{ hr}}{60 \text{ min}}\right)\left(\frac{1 \text{ day}}{24 \text{ hr}}\right)$$
$$= 22{,}978 \text{ ft}^3$$

The depth of each tank is

$$d = \frac{V}{A} = \frac{4V}{\pi D^2}$$
$$= \frac{(4)(22{,}978 \text{ ft}^3)}{\pi (70 \text{ ft})^2}$$
$$= 5.97 \text{ ft} \quad (6 \text{ ft})$$

The answer is (B).

3. *Salmonella typhi*, *Giardia lamblia*, *Poliovirus*, and *Endamoeba histolytica* (organisms I, III, V, and VII) are all pathogens that can be transmitted through the water supply and wastewater.

Escherichia coli (organism II) is a bacteria commonly present in wastewater and is used as an indicator organism, but it is not typically considered a pathogen. However, *E. Coli* can cause gastroenteritis (diarrhea).

Sphaerotilis natans (organism IV) is a filamentous form of bacteria that can adversely affect operation of the activated sludge process, but it is not pathogenic.

Ceriodaphnia dubia (organism VI) is the daphnid shrimp used in toxicity testing. It is not a pathogen.

Pimephales promelas (organism VIII) is the fathead minnow used in toxicity testing. It is not a pathogen.

The answer is (B).

4. The control survival is high for all of the species tested, so all results can be considered valid. The bioassay report would otherwise indicate significant control problems. The species with the lowest percent effluent resulting in no observed effect concentration (NOEC) is *Ceriodaphnia dubia* with a NOEC of 2.0% effluent. A chronic toxicity unit is the reciprocal of the effluent dilution that caused no unacceptable effect on the test organisms by the end of the chronic exposure period.

The chronic toxicity is

$$\text{TU}_c = \frac{100\%}{\text{NOEC}} = \frac{100\%}{2.0\%}$$
$$= 50 \text{ chronic toxicity units}$$

For chronic protection, the CCC must not exceed 1.0 chronic toxicity unit. The CCC is the chronic toxicity divided by the chronic initial dilution (CID), which is the dilution in the mixing zone.

$$\text{CCC} = \frac{\text{TU}_c}{\text{CID}} = \frac{50 \text{ chronic toxicity units}}{100 \text{ dilution}}$$
$$= 0.5 \text{ chronic toxicity unit}$$

The CCC is less than 1.0 chronic toxicity unit.

The answer is (C).

5. Statement I is false. Zone 1 would be characterized as mesosaprobic and representative of diluted wastewater, an inadequately treated effluent, or a degraded stream. Typical of a degraded water body in transition, this zone reveals the presence of some DO, but at a low level (4.5 mg) for natural water; a somewhat elevated level of BOD (3 mg/L); and high total coliform counts.

Statement IV is false. Zone 1 would be characterized as mesosaprobic and represents a degraded stream in transition. Zone 3 is highly polluted water that would be characterized as polysaprobic.

Statements II and III are true.

The answer is (C).

6. Twenty percent of the delivered solid waste (5% moisture and 15% classified materials) is removed prior to entering the furnace. The balance is 180 tons/day per power module. There are three power modules.

The solid waste brought to the plant is

$$W_{\text{waste}} = \frac{NC}{1 - F}$$

N is the number of power modules, C is the input rate per module, and F is the fraction of delivered solid waste that is removed before entering the furnace.

$$W_{\text{waste}} = \frac{(3 \text{ modules}) \left(180 \, \dfrac{\text{tons}}{\text{day-module}} \right)}{1 - 0.20}$$
$$= 675 \text{ tons/day}$$

The population is

$$P = \frac{m}{P_e} = \frac{\left(675 \, \dfrac{\text{tons}}{\text{day}} \right) \left(2000 \, \dfrac{\text{lbm}}{\text{ton}} \right)}{7 \, \dfrac{\text{lbm}}{\text{person-day}}}$$
$$= 1.93 \times 10^5 \text{ people} \quad (200{,}000 \text{ people})$$

The answer is (B).

7. The total mass of solid waste generated in 1 yr is

$$m = NP_e D$$

N is the number of people, P_e is the mass of solid waste generated per person per day, and D is the duration in days.

$$m = (25{,}000 \text{ people}) \left(8 \, \dfrac{\text{lbm}}{\text{person-day}} \right) \left(365 \, \dfrac{\text{days}}{\text{yr}} \right)$$
$$= 73{,}000{,}000 \text{ lbm/yr}$$

The annual in-place volume of solid waste is

$$V_{sw} = \frac{m}{\rho} = \frac{73{,}000{,}000\ \dfrac{\text{lbm}}{\text{yr}}}{1200\ \dfrac{\text{lbm}}{\text{yd}^3}}$$

$$= 60{,}833\ \text{yd}^3$$

The annual soil volume required is

$$V_{soil} = V_{sw}R$$

R is the ratio of soil to solid waste.

$$V_{soil} = (60{,}833\ \text{yd}^3)\left(\frac{1\ \text{yd}^3\ \text{soil}}{4\ \text{yd}^3\ \text{solid waste}}\right)$$

$$= 15{,}208\ \text{yd}^3$$

The annual in-place mass of soil required is

$$m_{soil} = V_{soil}\rho$$

ρ is the density of soil.

$$m_{soil} = (15{,}208\ \text{yd}^3)\left(130\ \frac{\text{lbm}}{\text{ft}^3}\right)$$

$$\times \left(27\ \frac{\text{ft}^3}{\text{yd}^3}\right)\left(\frac{1\ \text{ton}}{2000\ \text{lbm}}\right)$$

$$= 26{,}690\ \text{tons}\quad(25{,}000\ \text{tons})$$

The answer is (B).

8. The gross velocity of groundwater flow is

$$v_{gross} = Ki = K\frac{\Delta H}{L}$$

$$= \left(0.1\ \frac{\text{m}}{\text{d}}\right)\left(\frac{40\ \text{m} - 0\ \text{m}}{2000\ \text{m}}\right)$$

$$= 2 \times 10^{-3}\ \text{m/d}$$

The pore velocity of groundwater flow is

$$v_{pore} = \frac{v_{gross}}{n}$$

n represents porosity.

$$v_{pore} = \frac{2 \times 10^{-3}\ \dfrac{\text{m}}{\text{d}}}{0.25}$$

$$= 8 \times 10^{-3}\ \text{m/d}$$

The coefficient of retardation, C_r, is the ratio of the velocity of the centroid of the contaminant plume to the groundwater velocity. The velocity of the contaminant plume is

$$v_{plume} = C_r v_{pore}$$

Try the coefficient of retardation for more than 640 d travel time. The velocity of the plume is

$$v_{plume} = C_r v_{pore}$$

$$= (0.17)\left(8.0 \times 10^{-3}\ \frac{\text{m}}{\text{d}}\right)$$

$$= 1.36 \times 10^{-3}\ \text{m/d}$$

The travel time is

$$t = \frac{L}{v_{plume}}$$

$$= \frac{2000\ \text{m}}{1.36 \times 10^{-3}\ \dfrac{\text{m}}{\text{d}}}$$

$$= 1.47 \times 10^6\ \text{d}$$

The result that shows the coefficient of retardation used in the preceding equation is appropriate, because the travel time is more than 640 d.

$$t = (1.47 \times 10^6\ \text{d})\left(\frac{1\ \text{yr}}{365\ \text{d}}\right)$$

$$= 4029\ \text{yr}\quad(4000\ \text{yr})$$

The answer is (A).

9. Coulomb's equation for active force on soil should be used to solve this problem.

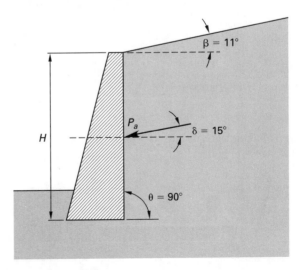

θ is the angle between a horizontal plane and the interior face of the retaining wall. Since the retaining wall surface in contact with the soil is vertical, θ is $90°$.

$$k_a = \frac{\sin^2(\phi + \theta)}{\sin^2 \theta \sin(\theta - \delta) \left(1 + \sqrt{\dfrac{\sin(\delta + \phi)\sin(\phi - \beta)}{\sin(\theta - \delta)\sin(\beta + \theta)}}\right)^2}$$

$$= \frac{\sin^2(30° + 90°)}{\sin^2 90° \sin(90° - 15°)}$$

$$\times \left(1 + \sqrt{\frac{\sin(15° + 30°)\sin(30° - 11°)}{\sin(90° - 15°)\sin(11° + 90°)}}\right)^2$$

$$= 0.348$$

The active force per unit length of wall is

$$P_a = \tfrac{1}{2}\gamma H^2 k_a$$
$$= \left(\frac{1}{2}\right)\left(18 \ \frac{\text{kN}}{\text{m}^3}\right)(4.0 \text{ m})^2 (0.348)$$
$$= 50.1 \text{ kN/m} \quad (50 \text{ kN/m})$$

The answer is (B).

10. Before the sand fill is placed on top of the silt deposit, the total stress, water pressure, and effective stress are, respectively,

$$\sigma = \gamma_{\text{sat}} z_A = \left(18 \ \frac{\text{kN}}{\text{m}^3}\right)(11.0 \text{ m})$$
$$= 198 \text{ kN/m}^2$$
$$u = \gamma_w z_w = \left(9.81 \ \frac{\text{kN}}{\text{m}^3}\right)(11.0 \text{ m} - 1.0 \text{ m})$$
$$= 98.1 \text{ kN/m}^2$$
$$\sigma' = \sigma - u = 198 \ \frac{\text{kN}}{\text{m}^2} - 98.1 \ \frac{\text{kN}}{\text{m}^2}$$
$$= 99.9 \text{ kN/m}^2$$

After placing the sand fill on top of the silt deposit, the total stress becomes

$$\sigma_{\text{filled}} = \left(18 \ \frac{\text{kN}}{\text{m}^3}\right)(11.0 \text{ m}) + \left(20 \ \frac{\text{kN}}{\text{m}^3}\right)(5.0 \text{ m})$$
$$= 298 \text{ kN/m}^2$$

At the end of the consolidation process, the water pressure is the same as before the placement of the fill.

$$u = 98.1 \text{ kN/m}^2$$
$$\sigma'_{\text{filled}} = \sigma_{\text{filled}} - u = 298 \ \frac{\text{kN}}{\text{m}^2} - 98.1 \ \frac{\text{kN}}{\text{m}^2}$$
$$= 199.9 \text{ kN/m}^2$$

Therefore, the increase in effective stress at point A is

$$\Delta\sigma'_A = \sigma'_{\text{filled}} - \sigma' = 199.9 \ \frac{\text{kN}}{\text{m}^2} - 99.9 \ \frac{\text{kN}}{\text{m}^2}$$
$$= 100 \text{ kN/m}^2 \quad (100 \text{ kPa})$$

The answer is (C).

11. A properly drawn flow net should be similar to the one shown, or at least have approximately the same N_f-to-N_p ratio. Do not spend a lot of time trying to draw a perfect flow net.

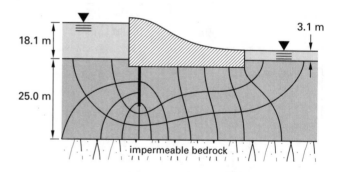

The flow net shows three flow tubes ($N_f = 3$) and 11 head drops ($N_p = 11$). The entire flow is

$$Q = KH\frac{N_f}{N_p}L$$
$$= \left(2 \times 10^{-6} \ \frac{\text{m}}{\text{s}}\right)(18.1 \text{ m} - 3.1 \text{ m})\left(\frac{3}{11}\right)(300 \text{ m})$$
$$= 2.45 \times 10^{-3} \text{ m}^3/\text{s} \quad (2.5 \times 10^{-3} \text{ m}^3/\text{s})$$

The answer is (D).

12. The Boussinesq stress contour chart may be used to solve this problem.

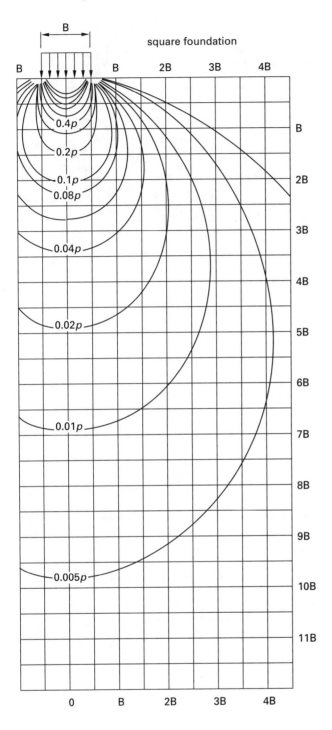

square foundation

For a column load, N, of 800 kN and a bearing pressure, p, of 400 kN/m², the width of the square footing is

$$B = \sqrt{\frac{N}{p}} = \sqrt{\frac{800 \text{ kN}}{400 \frac{\text{kN}}{\text{m}^2}}}$$

$$= 1.414 \text{ m}$$

Therefore,

$$\frac{y}{B} = \frac{3 \text{ m}}{1.414 \text{ m}}$$

$$= 2.12$$

The resulting factor from the Boussinesq chart is $0.095p$. The resulting stress is

$$\sigma_y = 0.095p = (0.095) \left(400 \frac{\text{kN}}{\text{m}^2} \right)$$

$$= 38 \text{ kN/m}^2$$

For the 100 kN/m² foundation bearing pressure,

$$B = \sqrt{\frac{800 \text{ kN}}{100 \frac{\text{kN}}{\text{m}^2}}}$$

$$= 2.83 \text{ m}$$

Therefore,

$$\frac{y}{B} = \frac{3 \text{ m}}{2.83 \text{ m}}$$

$$= 1.06$$

The resulting factor from the Boussinesq stress contour chart is $0.3p$. The resulting stress is

$$\sigma_y = 0.3p = (0.3) \left(100 \frac{\text{kN}}{\text{m}^2} \right)$$

$$= 30 \text{ kN/m}^2$$

The change in stress at 3 m below the center of the foundation is

$$\Delta \sigma_y = 30 \frac{\text{kN}}{\text{m}^2} - 38 \frac{\text{kN}}{\text{m}^2}$$

$$= -8 \text{ kN/m}^2 \quad (-10 \text{ kPa})$$

The answer is (B).

13.

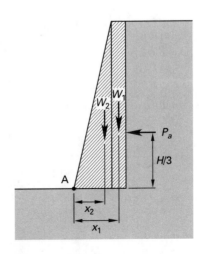

In this case, the only force that tends to resist overturning is the weight of the wall, W. The only force that tends to overturn the wall is the active force, P_a.

The wall's angle of friction, δ, is $0°$, so the active force can be calculated using the Rankine theory.

$$
\begin{aligned}
P_a &= \tfrac{1}{2}\gamma H^2 k_a \\
&= \left(\frac{1}{2}\right)\left(19\ \frac{kN}{m^3}\right)(5.0\ m)^2 (0.32) \\
&= 76\ kN/m
\end{aligned}
$$

The active earth pressure coefficient is

$$
\begin{aligned}
k_a &= \tan^2\left(45° - \frac{\phi}{2}\right) = \tan^2\left(45° - \frac{31°}{2}\right) \\
&= 0.32
\end{aligned}
$$

The overturning moment per unit length of wall is

$$
\begin{aligned}
\sum M_{\text{overturning}} &= P_a \frac{H}{3} = \left(76\ \frac{kN}{m}\right)\left(\frac{5.0\ m}{3}\right) \\
&= 126.7\ kN\cdot m/m
\end{aligned}
$$

The weight of the wall, W, is equal to W_1 plus W_2.

$$
\begin{aligned}
W_1 &= HD\gamma_{\text{concrete}} \\
&= (5.0\ m)(1.0\ m)\left(24\ \frac{kN}{m^3}\right) \\
&= 120\ kN/m \\
W_2 &= \tfrac{1}{2}HD\gamma_{\text{concrete}} \\
&= \left(\frac{1}{2}\right)(5.0\ m)(1.5\ m)\left(24\ \frac{kN}{m^3}\right) \\
&= 90\ kN/m
\end{aligned}
$$

The resisting moment per unit length of wall is

$$
\begin{aligned}
\sum M_{\text{resisting}} &= W_1 x_1 + W_2 x_2 \\
&= \left(120\ \frac{kN}{m}\right)(2.0\ m) + \left(90\ \frac{kN}{m}\right)(1.0\ m) \\
&= 330\ kN\cdot m/m
\end{aligned}
$$

The factor of safety against overturning about the toe (point A of the illustration) can be expressed as

$$
\begin{aligned}
\text{FS}_{\text{OT}} &= \frac{\sum M_{\text{resisting}}}{\sum M_{\text{overturning}}} = \frac{330\ \dfrac{kN\cdot m}{m}}{126.7\ \dfrac{kN\cdot m}{m}} \\
&= 2.6
\end{aligned}
$$

The answer is (C).

14. Since ϕ is $0°$, the Taylor slope stability chart can be used to solve this problem.

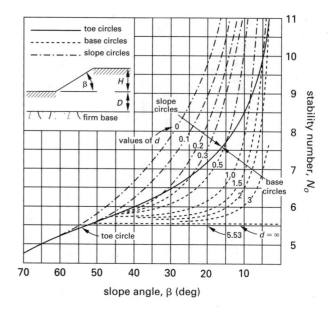

Source: *Soil Mechanics*, NAVFAC Design Manual DM-7.1, May 1982.

The depth factor, d, is the ratio of the vertical distance between the toe of the slope and the firm base below the clay layer, D, to the slope height, H (the depth of the cut).

$$
\begin{aligned}
d &= \frac{D}{H} = \frac{3.0\ m}{6.0\ m} \\
&= 0.5
\end{aligned}
$$

The Taylor chart shows that, for a depth factor of 0.5 and a slope angle, β, of $55°$, the stability number, N_0, is 5.5.

The cohesive factor of safety for the slope is

$$
\begin{aligned}
F_{\text{cohesive}} &= \frac{N_0 c}{\gamma H} \\
&= \frac{(5.5)(50\ kPa)}{\left(19\ \dfrac{kN}{m^3}\right)(6.0\ m)} \\
&= 2.41 \quad (2.5)
\end{aligned}
$$

The answer is (D).

15. A strength envelope can be constructed using the data given. Mohr's circle can then be constructed tangent to the strength envelope at point 2, as shown. From this arrangement, the major and minor principal stresses are found.

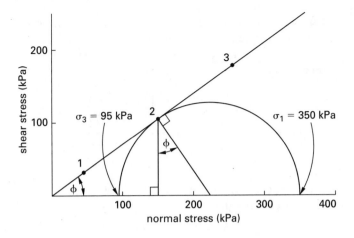

The answer is (B).

16. This problem requires use of the American Association of State Highway and Transportation Officials (AASHTO) classification table for soils and soil-aggregate mixtures.

The percentages of material passing the no. 10, no. 40, and no. 200 sieves are

no. 10 (2.00 mm)	45%	
no. 40 (0.425 mm)	35%	
no. 200 (0.075 mm)	28%	

The liquid limit (LL) and plasticity index (PI) of the material passing the no. 40 sieve are 34 and 13, respectively. Therefore, working from left to right on the AASHTO classification chart, the appropriate subgroup for this soil is A-2-6.

The AASHTO classification also contains a number in parentheses. This number is the group index. For subgroups A-2-6 and A-2-7, the group index is

$$I_g = 0.01(F_{200} - 15)(PI - 10)$$

F_{200} is the percentage of soil passing the no. 200 sieve.

The group index is always reported to the nearest whole number. (If it is calculated to be negative, then it is reported as zero.)

$$\begin{aligned} I_g &= 0.01(F_{200} - 15)(PI - 10) \\ &= (0.01)(28 - 15)(13 - 10) \\ &= 0.39 \quad (0) \end{aligned}$$

Therefore, the final AASHTO classification for this soil is A-2-6 (0).

The answer is (A).

17. The resultant is a 32 kips force located $\bar{x}$ from the leftmost load.

$$\begin{aligned} \bar{x} &= \frac{\sum P_i x_i}{\sum P_i} \\ &= \frac{(8 \text{ kips})(0 \text{ ft}) + (8 \text{ kips})(10 \text{ ft}) + (16 \text{ kips})(22 \text{ ft})}{8 \text{ kips} + 8 \text{ kips} + 16 \text{ kips}} \\ &= 13.5 \text{ ft} \end{aligned}$$

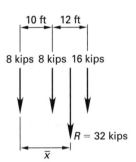

The absolute maximum bending moment occurs under the 16 kips load when the distance from midspan to 16 kips is equidistant from midspan to resultant. Thus, the distance from midspan to 16 kips is

$$\frac{22 \text{ ft} - 13.5 \text{ ft}}{2} = 4.25 \text{ ft}$$

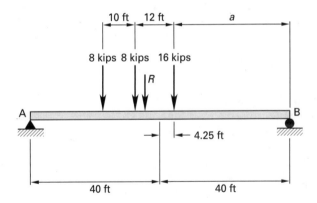

The reaction at the right support is found by summing moments about the left support.

$$\begin{aligned} \sum M_A &= 0 = (R_B)(80 \text{ ft}) - (32 \text{ kips})(40 \text{ ft} - 4.25 \text{ ft}) \\ R_B &= 14.3 \text{ kips} \end{aligned}$$

The bending moment under the 16 kips load is

$$\begin{aligned} M_{max} &= R_B a \\ &= (14.3 \text{ kips})(40 \text{ ft} - 4.25 \text{ ft}) \\ &= 511 \text{ ft-kips} \quad (510 \text{ ft-kips}) \end{aligned}$$

The answer is (A).

18.

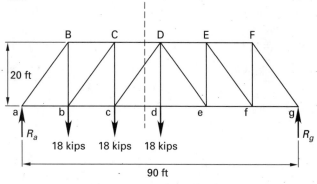

Sum moments about point g. (Assume clockwise moment is positive.)

$$\sum M_g = \sum_i r_i F_i = 0$$

$$(90 \text{ ft})R_a - (18 \text{ kips})(75 \text{ ft} + 60 \text{ ft} + 45 \text{ ft}) = 0$$

$$R_a = 36 \text{ kips}$$

Pass a cutting plane vertically between panel points C and D; consider the left side as the free body. Take moments about point c.

$$\sum M_c = \sum_i r_i F_i = 0$$

$$(20 \text{ ft})F_{CD} + (36 \text{ kips})(30 \text{ ft}) - (18 \text{ kips})(15 \text{ ft}) = 0$$

$$F_{CD} = -40.5 \text{ kips} \quad (-40 \text{ kips})$$

The answer is (C).

19. Since the resultant compression force acts within the middle one-third of the cross section, the gross cross-sectional area is effective in resisting the combined axial and bending stresses.

$$A = bh = (32 \text{ in})(32 \text{ in})$$
$$= 1024 \text{ in}^2$$

$$I = \frac{bh^3}{12} = \frac{(32 \text{ in})(32 \text{ in})^3}{12}$$
$$= 87{,}381 \text{ in}^4$$

P is the axial compression force, e is the eccentricity, and c is the distance from the center of the column to the edge.

$$f_c = -\frac{P}{A} - \frac{Pec}{I}$$
$$= -\frac{110 \text{ kips}}{1024 \text{ in}^2} - \frac{(110 \text{ kips})(5 \text{ in})(16 \text{ in})}{87{,}381 \text{ in}^4}$$
$$= -0.208 \text{ kips/in}^2 \quad (0.2 \text{ kips/in}^2)$$

The answer is (A).

20. The area of one no. 10 bar is 1.27 in². The yield stress of a grade 60 rebar is 60,000 lbf/in². The total compression area required to balance the tension in the rebars is

$$A_c = \frac{A_s f_y}{0.85 f_c'} = \frac{(4)(1.27 \text{ in}^2)\left(60{,}000 \, \dfrac{\text{lbf}}{\text{in}^2}\right)}{(0.85)\left(4000 \, \dfrac{\text{lbf}}{\text{in}^2}\right)}$$

$$= 89.6 \text{ in}^2$$

This is greater than the area of the "ears" (i.e., of the sides of the trough).

$$A_{\text{top}} = (2)(5 \text{ in})(4 \text{ in})$$
$$= 40 \text{ in}^2$$

Therefore, the compression zone extends below the trough.

$$a_w = \frac{A_c - A_{\text{top}}}{b} = \frac{89.6 \text{ in}^2 - 40 \text{ in}^2}{22 \text{ in}}$$
$$= 2.25 \text{ in}$$

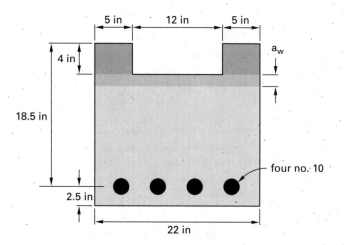

The design moment strength of the section is ϕM_n. Taking the moments of the internal compression force about the centroid of tension steel gives

$$\phi M_n = \sum \phi C_{ci}(d - \lambda_i)$$

$$= (0.9) \begin{pmatrix} (0.85)\left(4 \, \dfrac{\text{kips}}{\text{in}^2}\right)(40 \text{ in}^2)(18.5 \text{ in} - 2 \text{ in}) \\ + (0.85)\left(4 \, \dfrac{\text{kips}}{\text{in}^2}\right)(22 \text{ in})(2.25 \text{ in}) \\ \times \left(18.5 \text{ in} - 4 \text{ in} - \dfrac{2.25 \text{ in}}{2}\right) \end{pmatrix}$$

$$\times \left(\frac{1 \text{ ft}}{12 \text{ in}}\right)$$

$$= 337 \text{ ft-kips} \quad (340 \text{ ft-kips})$$

The answer is (B).

SOLUTIONS — MORNING SESSION **75**

21. Allowing 3 in minimum cover plus 1 in bar diameter for the separation between the no. 8 rebars in each direction, the effective footing depth is 16 in.

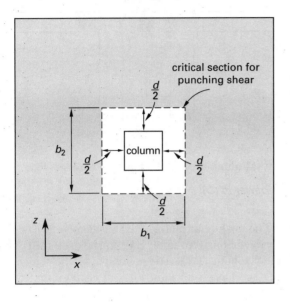

The critical punching shear perimeter is at $d/2$ from the column face. Thus,

$$b_o = 4(h + d) = (4)(14 \text{ in} + 16 \text{ in})$$
$$= 120 \text{ in}$$

For square columns, the controlling punching shear capacity is given in ACI 318.

$$\phi V_n = 4\phi\sqrt{f'_c}b_o d$$
$$= (4)(0.75)\sqrt{3000 \ \frac{\text{lbf}}{\text{in}^2}}(120 \text{ in})(16 \text{ in})$$
$$= 315{,}488 \text{ lbf} \quad (316 \text{ kips})$$

The design force on the column is uniformly distributed over the 8.5 ft × 8.5 ft spread footing. The punching shear excludes the distributed force within the distance d to every side of the column face. Thus,

$$V_u = w_u A_{\text{trib}} \leq \phi V_n$$

$$w_u \left((8.5 \text{ ft})(8.5 \text{ ft}) - \left(\frac{14 \text{ in} + (2)(16 \text{ in})}{12 \ \frac{\text{in}}{\text{ft}}} \right)^2 \right)$$

$$\leq 316 \text{ kips}$$
$$w_u = 5.49 \text{ kips/ft}^2$$
$$P_u \leq w_u A_f \leq \left(5.49 \ \frac{\text{kips}}{\text{ft}^2} \right)(8.5 \text{ ft})(8.5 \text{ ft})$$
$$\leq 396.7 \text{ kips} \quad (400 \text{ kips})$$

The answer is (B).

22. The location of the centroid of the fastener group is known from symmetry. The polar moment of inertia of the group is

$$I_p = I_x + I_y$$
$$= (4)\left((3 \text{ in})^2 + (6 \text{ in})^2\right) + (10)\left(\frac{5 \text{ in}}{2}\right)^2$$
$$= 242.5 \text{ in}^2$$

The statically equivalent force system is 36 kips upward through the centroid plus a counterclockwise moment of (8.5 in)(36 kips) = 306 in-kips. The critical fastener is an extreme fastener for which the y-component acts down and adds algebraically to the y-component of the direct force. Thus,

$$R_{py} = \frac{P}{n} = \frac{36 \text{ kips}}{10}$$
$$= 3.6 \text{ kips}$$
$$R_{my} = \frac{Mx}{I_p} = \frac{(306 \text{ in-kips})(2.5 \text{ in})}{242.5 \text{ in}^2}$$
$$= 3.15 \text{ kips}$$
$$R_{mx} = \frac{My}{I_p} = \frac{(306 \text{ in-kips})(6 \text{ in})}{242.5 \text{ in}^2}$$
$$= 7.57 \text{ kips}$$

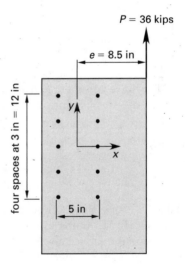

The resultant is found by vector addition of the components.

$$R = \sqrt{R_{mx}^2 + (R_{my} + R_{py})^2}$$
$$= \sqrt{(7.57 \text{ kips})^2 + (3.15 \text{ kips} + 3.6 \text{ kips})^2}$$
$$= 10.1 \text{ kips} \quad (10 \text{ kips})$$

The answer is (C).

PROFESSIONAL PUBLICATIONS, INC.

23. The flexible diaphragm behaves as a simple beam between the supporting shearwalls. The maximum bending moment occurs in the 80 ft span between shearwalls on lines B and C.

$$M_{\max} = \frac{wL^2}{8} = \frac{\left(320 \ \frac{\text{lbf}}{\text{ft}}\right)(80 \ \text{ft})^2}{8}$$
$$= 256{,}000 \ \text{ft-lbf}$$

The moment is resisted by the chords, which are nominally 55 ft apart. Thus,

$$C = T = \frac{M_{\max}}{b}$$
$$= \frac{256{,}000 \ \text{ft-lbf}}{55 \ \text{ft}}$$
$$= 4654 \ \text{lbf} \quad (4700 \ \text{lbf})$$

The answer is (B).

24. The Mohr's circle drawn for the state of stress on any two orthogonal faces plots as a point at $(\sigma, 0)$. Thus, the radii of the stress circles are all zero, which means that the shear stress is zero for all possible orientations of the element. Hence, there can be no shear slip in the element. Plastic deformation can occur in steel only in the presence of shear stresses; therefore, ductile behavior cannot occur. Since the shear stress is zero under this triaxial state of stress, failure can occur only by cleavage.

The answer is (C).

25. Find the LOS using the LOS criteria tables and similar tools in the HCM. At the given free-flow speed of 65 mph, the maximum service flow rate (1350 pcphpl) is greater than the maximum service flow rate for LOS B (1170 pcphpl) and less than the maximum service flow rate for LOS C (1680 pcphpl). Therefore the freeway is considered to be operating at LOS C.

The answer is (B).

26. In a three-leg intersection, there are three conflict points for each of the conflict types (crossing, merging, and diverging).

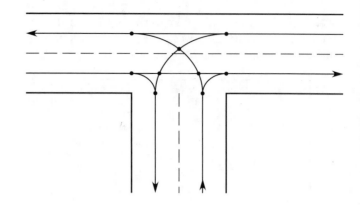

The total number of conflict points is nine.

The answer is (C).

27. The peak hour factor (PHF) is defined as the ratio of the total hourly volume (HV) to the peak rate of flow within the hour (PV). The volume of traffic that passes during the busiest 15 min, $V_{15 \ \text{min}}$, is 750 vph.

$$\text{PHF} = \frac{\text{HV}}{\text{PV}} = \frac{\text{HV}}{4V_{15 \ \text{min}}}$$
$$= \frac{\left(2400 \ \frac{\text{veh}}{\text{hr}}\right)}{(4)\left(750 \ \frac{\text{veh}}{\text{hr}}\right)}$$
$$= 0.80$$

The answer is (B).

28. First determine the density, D, from the given equation using the given average travel speed of 50 mph.

$$S = 65 \ \frac{\text{mi}}{\text{hr}} - \left(0.42 \ \frac{\frac{\text{mi}}{\text{hr}}}{\frac{\text{veh}}{\text{mi}}}\right) D$$

$$50 \ \frac{\text{mi}}{\text{hr}} = 65 \ \frac{\text{mi}}{\text{hr}} - \left(0.42 \ \frac{\frac{\text{mi}}{\text{hr}}}{\frac{\text{veh}}{\text{mi}}}\right) D$$

$$D = 35.7 \ \text{vpm}$$

The flow rate can be calculated using an HCM traffic density equation, given the speed and density.

$$D = \frac{\text{v}}{S}$$
$$35.7 \ \frac{\text{veh}}{\text{mi}} = \frac{\text{v}}{50 \ \frac{\text{mi}}{\text{hr}}}$$
$$\text{v} = 1785 \ \text{vph} \quad (1800 \ \text{vph})$$

The answer is (B).

29. Option (A) is false. The average daily traffic (ADT) is the average of 24 hr counts collected over a number of days greater than one but less than 365. The given definition is for the annual average daily traffic (AADT).

Option (B) is false, because traffic control devices such as traffic lights and stop signs usually cause fixed delay on roadways. The causes of operational delay include traffic side frictions.

Option (C) is true. Space mean speed is always less than time mean speed.

Option (D) is false. The first part of the statement is correct (local roads have more access than mobility), but the second part of the statement is false because local roads nationwide carry a much smaller fraction (only about 20%) of the travel volume. Overall, the statement should be considered false.

The answer is (C).

30. The stopping sight distance, S, is the sum of brake reaction and brake distance. From the AASHTO Green Book, the brake reaction for 45 mph speed is 165.4 ft with 2.5 sec perception reaction time.

The brake distance for a highway on grade, G, is also given by the AASHTO Green Book. The deceleration, a, is assumed to be 11.2 ft/sec^2 as a default value.

$$d = \frac{\text{v}^2}{30\left(\dfrac{a}{g} \pm G\right)}$$

$$= \frac{\left(45\ \dfrac{\text{m}}{\text{hr}}\right)^2}{(30)\left(\dfrac{11.2\ \dfrac{\text{ft}}{\text{sec}^2}}{32.2\ \dfrac{\text{ft}}{\text{sec}^2}} - 0.05\right)}$$

$$= 226.6\ \text{ft}$$

$$S = 165.4\ \text{ft} + 226.6\ \text{ft}$$
$$= 392\ \text{ft} \quad (390\ \text{ft})$$

The answer can also be obtained by interpolation from exhibits from the AASHTO Green Book.

The answer is (C).

31. Based on the given traffic volumes, link AB represents a major highway, and link CD is a minor road. The traffic is predominantly from A to B and B to A. The turning movements can be accommodated by at-grade turns. A diamond interchange can handle the large volumes between A and B and can incorporate the at-grade turns. A simple diamond interchange is the most suitable.

The answer is (C).

32. Use the average end area method to calculate the volume of cut and the volume of fill.

$$V = \frac{L(A_1 + A_2)}{2}$$

$$V_{\text{cut}} = \left(\frac{(100\ \text{ft})(9\ \text{ft}^2 + 15\ \text{ft}^2)}{2}\right)\left(\frac{1\ \text{yd}^3}{27\ \text{ft}^3}\right)$$

$$= 44.44\ \text{yd}^3$$

$$V_{\text{fill}} = \left(\frac{(100\ \text{ft})(24\ \text{ft}^2 + 32\ \text{ft}^2)}{2}\right)\left(\frac{1\ \text{yd}^3}{27\ \text{ft}^3}\right)$$

$$= 103.70\ \text{yd}^3$$

The net earth volume (NEV) is

$$\text{NEV} = V_{\text{cut}} - V_{\text{fill}}$$
$$= 44.44\ \text{yd}^3 - 103.70\ \text{yd}^3$$
$$= -59.26\ \text{yd}^3 \quad (60\ \text{yd}^3\ \text{of fill})$$

The answer is (B).

33. The unit volume produced from 1.0 in runoff from the watershed is

$$V_{\text{unit}} = A_d h$$

$$= \frac{(43\ \text{mi}^2)\left(640\ \dfrac{\text{ac}}{\text{mi}^2}\right)(1\ \text{in})}{12\ \dfrac{\text{in}}{\text{ft}}}$$

$$= 2293\ \text{ac-ft}$$

Since the actual flood hydrograph volume is 3260 ac-ft, the depth of runoff in the flood hydrograph is

$$d = \frac{V_{\text{actual}}}{V_{\text{unit}}} = \frac{3260\ \text{ac-ft}}{2293\ \text{ac-ft}}$$
$$= 1.42\ \text{in}$$

The actual flood hydrograph peak is 9300 ft^3/sec, so the unit hydrograph peak is

$$Q_{p,\text{unit}} = \frac{Q_{p,\text{actual}}}{d} = \frac{9300\ \dfrac{\text{ft}^3}{\text{sec}}}{1.42\ \text{in}}$$

$$= 6549\ \text{ft}^3/\text{in-sec} \quad (6500\ \text{ft}^3/\text{in-sec})$$

The answer is (D).

34. The coefficient of velocity is the ratio of actual velocity to the theoretical velocity at the nozzle.

$$C_\text{v} = \frac{\text{v}_{\text{actual}}}{\text{v}_{\text{th}}}$$

The theoretical velocity is

$$v_{th} = \frac{v_{actual}}{C_v} = v_2$$

$$= \frac{98 \ \frac{ft}{sec}}{0.95}$$

$$= 103.2 \ ft/sec$$

The ratio of the *vena contracta* (jet) area to the area of the nozzle is the coefficient of contraction.

$$C_c = \frac{A_{jet}}{A_2} = 0.8$$

$$A_{jet} = 0.8A_2$$

From the continuity equation,

$$Q_1 = Q_2$$

$$v_1 A_1 = v_2 A_{jet}$$

$$v_1 = \frac{0.8 A_2 v_2}{A_1} = \frac{0.8 D_2^2 v_2}{D_1^2}$$

$$= \frac{(0.8)(0.8 \ in)^2 \left(103.2 \ \frac{ft}{sec}\right)}{(3.15 \ in)^2}$$

$$= 5.33 \ ft/sec$$

The Bernouli equation from point 1 (base) to point 2 (jet) is

$$\frac{p_1 g_c}{\rho g} + z_1 + \frac{v_1^2}{2g} = \frac{p_2 g_c}{\rho g} + z_2 + \frac{v_2^2}{2g}$$

At 160°F, the density of water, ρ, is 61.0 lbm/ft³. The pressure at point 1 is found from

$$\frac{p_1 \left(32.2 \ \frac{lbm\text{-}ft}{lbf\text{-}sec^2}\right)}{\left(61.0 \ \frac{lbm}{ft^3}\right) g} + 0 + \frac{\left(5.33 \ \frac{ft}{sec}\right)^2}{2g}$$

$$= 0 + 0 + \frac{\left(103.2 \ \frac{ft}{sec}\right)^2}{2g}$$

$$p_1 = 10,061 \ lbf/ft^2$$

$$= \left(10,061 \ \frac{lbf}{ft^2}\right) \left(\frac{1 \ ft^2}{144 \ in^2}\right)$$

$$= 69.9 \ lbf/in^2 \quad (70 \ psi)$$

The answer is (C).

35. The Manning coefficient, n, for brick is 0.016. The cross-sectional area for a rectangular channel is

$$A = dw = (5 \ ft)(10 \ ft)$$

$$= 50 \ ft^2$$

The wetted perimeter is

$$P = 2d + w = (2)(5 \ ft) + 10 \ ft$$

$$= 20 \ ft$$

The hydraulic radius is

$$R = \frac{A}{P} = \frac{50 \ ft^2}{20 \ ft}$$

$$= 2.5 \ ft$$

The slope is

$$S = \left(\frac{Qn}{1.49 A R^{2/3}}\right)^2$$

$$= \left(\frac{\left(1060 \ \frac{ft^3}{sec}\right)(0.016)}{(1.49)(50 \ ft^2)(2.5 \ ft)^{2/3}}\right)^2$$

$$= 0.01517 \quad (1.5\%)$$

The answer is (B).

36. The NRCS curve numbers are obtained from tables describing standard ground cover types.

unit	curve number
1	39
2	72
3	98

The weighted curve number for the basin is

$$CN_{wt} = \frac{CN_1 A_1 + CN_2 A_2 + CN_3 A_3}{A_1 + A_2 + A_3}$$

$$= \frac{\begin{array}{c}(39)(200,000 \ ft^2) + (72)(300,000 \ ft^2) \\ + (98)(180,000 \ ft^2)\end{array}}{200,000 \ ft^2 + 300,000 \ ft^2 + 180,000 \ ft^2}$$

$$= 69$$

The storage capacity is

$$S = \frac{1000}{CN_{wt}} - 10 = \frac{1000}{69} - 10$$

$$= 4.49 \ in \quad (4.5 \ in)$$

The answer is (D).

37. The volume of the basin is

$$V_{\text{basin}} = Qt = \left(3.5 \ \frac{\text{ft}^3}{\text{sec}}\right)(50 \ \text{sec})$$

$$= 175 \ \text{ft}^3$$

The power required is

$$P = \mu G^2 V_{\text{basin}}$$

$$= \left(2.050 \times 10^{-5} \ \frac{\text{lbf-sec}}{\text{ft}^2}\right)\left(850 \ \frac{\text{ft}}{\text{sec-ft}}\right)^2 (175 \ \text{ft}^3)$$

$$= 2592 \ \text{lbf-ft/sec}$$

Assume turbulent flow. The power required to drive the impeller is

$$P = \frac{N_p n^3 D^5 \rho}{g_c}$$

The impeller diameter is

$$D = \left(\frac{Pg_c}{N_p n^3 \rho}\right)^{1/5}$$

$$= \left(\frac{\left(2592 \ \frac{\text{lbf-ft}}{\text{sec}}\right)\left(32.2 \ \frac{\text{lbm-ft}}{\text{lbf-sec}^2}\right)}{(5.5)\left(3 \ \frac{\text{rev}}{\text{sec}}\right)^3\left(62.4 \ \frac{\text{lbm}}{\text{ft}^3}\right)}\right)^{1/5}$$

$$= 1.55 \ \text{ft}$$

Calculate the Reynolds number to check the turbulent flow assumption.

$$\text{Re} = \frac{D^2 n \rho}{\mu g_c}$$

$$= \frac{(1.55 \ \text{ft})^2 \left(3 \ \frac{\text{rev}}{\text{sec}}\right)\left(62.4 \ \frac{\text{lbm}}{\text{ft}^3}\right)}{\left(2.050 \times 10^{-5} \ \frac{\text{lbf-sec}}{\text{ft}^2}\right)\left(32.2 \ \frac{\text{lbm-ft}}{\text{lbf-sec}^2}\right)}$$

$$= 681{,}333$$

The Reynolds number is greater than 10,000, so the assumption of turbulent flow is appropriate.

The answer is (B).

38. The basic reaction is

$$Al_2 (SO_4)_3 + 3Ca (OH)_2 \rightarrow 2Al(OH)_3 \downarrow + 3CaSO_4$$

The molecular weights of the reactants are as follows. For $Al_2(SO_4)_3$,

$$MW = \left(27 \ \frac{\text{g}}{\text{mol}}\right)(2) + \left(32 \ \frac{\text{g}}{\text{mol}} + \left(16 \ \frac{\text{g}}{\text{mol}}\right)(4)\right)(3)$$

$$= 342 \ \text{g/mol}$$

For $Ca(OH)_2$,

$$MW = 40 \ \frac{\text{g}}{\text{mol}} + \left(16 \ \frac{\text{g}}{\text{mol}} + 1 \ \frac{\text{g}}{\text{mol}}\right)(2)$$

$$= 74 \ \text{g/mol}$$

For $Al(OH)_3$,

$$MW = 27 \ \frac{\text{g}}{\text{mol}} + \left(16 \ \frac{\text{g}}{\text{mol}} + 1 \ \frac{\text{g}}{\text{mol}}\right)(3)$$

$$= 78 \ \text{g/mol}$$

For $CaSO_4$,

$$MW = 40 \ \frac{\text{g}}{\text{mol}} + 32 \ \frac{\text{g}}{\text{mol}} + \left(16 \ \frac{\text{g}}{\text{mol}}\right)(4)$$

$$= 136 \ \text{g/mol}$$

The alum dose is

$$D_{\text{alum}} = \left(342 \ \frac{\text{g}}{\text{mol alum}}\right)\left(\frac{1 \ \text{mol alum}}{2 \ \text{mol Al(OH)}_3}\right)$$

$$\times \left(\frac{1 \ \text{mol Al(OH)}_3}{78 \ \text{g Al(OH)}_3}\right)$$

$$\times \left(30 \ \frac{\text{mg}}{\text{L}} \ \text{Al(OH)}_3\right)$$

$$= 65.77 \ \text{mg/L}$$

The mass of alum required per day is

$$\dot{m} = D_{\text{alum}} Q$$

$$= \left(65.77 \ \frac{\text{mg}}{\text{L}}\right)\left(0.5 \ \frac{\text{m}^3}{\text{s}}\right)\left(1000 \ \frac{\text{L}}{\text{m}^3}\right)$$

$$\times \left(86{,}400 \ \frac{\text{s}}{\text{d}}\right)\left(\frac{1 \ \text{kg}}{10^6 \ \text{mg}}\right)$$

$$= 2841 \ \text{kg/d} \quad (3000 \ \text{kg/d})$$

The answer is (A).

39. For plain sedimentation, Type I settling occurs. Assume Stokes' law applies. The settling velocity is

$$v_s = \frac{(\text{SG}_{\text{particle}} - 1) \ D_{\text{ft}}^2 g}{18\nu}$$

$$= \frac{(2.2 - 1)\left(6.5 \times 10^{-5} \ \text{ft}\right)^2 \left(32.2 \ \frac{\text{ft}}{\text{sec}^2}\right)}{(18)\left(0.826 \times 10^{-5} \ \frac{\text{ft}^2}{\text{sec}}\right)}$$

$$= 0.0011 \ \text{ft/sec}$$

The Reynolds number is

$$\text{Re} = \frac{v_s D}{\nu} = \frac{\left(0.0011 \ \frac{\text{ft}}{\text{sec}}\right)\left(6.5 \times 10^{-5} \ \text{ft}\right)}{0.826 \times 10^{-5} \ \frac{\text{ft}^2}{\text{sec}}}$$

$$= 0.0086$$

Since the Reynolds number is less than 1, Stokes' law applies. For 100% removal, the settling velocity is the overflow rate (OFR).

$$\text{OFR} = 0.0011 \text{ ft}^3/\text{ft}^2\text{-sec}$$

The area of the tank is

$$A = \frac{Q}{\text{OFR}} = \frac{18 \dfrac{\text{ft}^3}{\text{sec}}}{0.0011 \dfrac{\text{ft}^3}{\text{ft}^2\text{-sec}}}$$
$$= 16{,}363 \text{ ft}^2 \quad (16{,}000 \text{ ft}^2)$$

The depth of the tank is

$$h = t_{\text{settling}} v_s$$
$$= (2.5 \text{ hr}) \left(3600 \frac{\text{sec}}{\text{hr}}\right) \left(0.0011 \frac{\text{ft}}{\text{sec}}\right)$$
$$= 9.9 \text{ ft} \quad (10 \text{ ft})$$

The answer is (D).

40. Statement II is false. Chlorine demand affects the dose required to achieve a particular concentration, but it does not directly relate to removal efficiency.

Statement IV is false. Filtration has an important effect on overall removal efficiency, and slow sand filtration is no exception.

Statements I and III are true.

The answer is (B).

Solutions
Environmental

---

41. The average daily flow is based on per capita factors in the absence of site specific data. Common values are 100–125 gal/capita-day.

The average daily domestic flow is

$$Q_{\text{dom}} = Q_{\text{cap}}P$$

Q_{cap} is the per capita daily domestic flow. P is the number of people (capita).

$$Q_{\text{dom}} = \left(125 \ \frac{\text{gal}}{\text{capita-day}}\right)(65{,}000 \ \text{capita})\left(\frac{1 \ \text{MG}}{10^6 \ \text{gal}}\right)$$
$$= 8.125 \ \text{MGD}$$

The industrial flow is

$$Q_{\text{ind}} = Q_{\text{unit}}A$$

Q_{unit} is the daily industrial flow per unit area. A is the area contributing to industrial flow.

$$Q_{\text{ind}} = \left(10{,}000 \ \frac{\text{gal}}{\text{ac-day}}\right)(200 \ \text{ac})\left(\frac{1 \ \text{MG}}{10^6 \ \text{gal}}\right)$$
$$= 2 \ \text{MGD}$$

The excess infiltration is

$$Q_{\text{inf}} = EdL$$

E is the excess infiltration rate, d is the diameter of pipe, and L is the length of pipe.

$$Q_{\text{inf}} = \left(600 \ \frac{\text{gal}}{\text{day-in-mi}}\right)\begin{pmatrix} (24 \ \text{in})(5 \ \text{mi}) \\ + (12 \ \text{in})(16 \ \text{mi}) \\ + (8 \ \text{in})(22 \ \text{mi}) \end{pmatrix}$$
$$\times \left(\frac{1 \ \text{MG}}{10^6 \ \text{gal}}\right)$$
$$= 0.293 \ \text{MGD}$$

The total annual average daily flow is

$$Q_{\text{total}} = Q_{\text{dom}} + Q_{\text{ind}} + Q_{\text{inf}}$$
$$= 8.125 \ \text{MGD} + 2.0 \ \text{MGD} + 0.293 \ \text{MGD}$$
$$= 10.418 \ \text{MGD}$$

The design peak daily flow is

$$Q_{\text{peak}} = Q_{\text{total}}\left(\frac{200\%}{100\%}\right)$$
$$= (10.418 \ \text{MGD})(2.0)$$
$$= 20.8 \ \text{MGD} \quad (20 \ \text{MGD})$$

The answer is (C).

42. The maximum daily design flow is

$$Q_{\text{max}} = Q_{\text{avg}}\text{PF}_{\text{d}}$$

Q_{avg} is the annual average flow rate in MGD. PF_{d} is the peaking factor for the maximum daily flow.

$$Q_{\text{max}} = (3.5 \ \text{MGD})(2.0)$$
$$= 7.0 \ \text{MGD}$$

The peak hour design flow is

$$Q_{\text{peak}} = Q_{\text{avg}}\text{PF}_{\text{hr}}$$

PF_{hr} is the peaking factor for maximum hourly flow.

$$Q_{\text{peak}} = (3.5 \ \text{MGD})(4.0)$$
$$= 14 \ \text{MGD}$$

The area of the clarifier is calculated by dividing the maximum daily flow by the overflow rate (OFR).

$$A = \frac{Q_{\text{max}}}{\text{OFR}} = \frac{(7.0 \ \text{MGD})\left(10^6 \ \frac{\text{gal}}{\text{MG}}\right)}{800 \ \frac{\text{gal}}{\text{day-ft}^2}}$$
$$= 8750 \ \text{ft}^2$$

The area based on the peak hour flow is

$$A = \frac{Q_{\text{peak}}}{O_{\text{OFR}}} = \frac{(14.0 \ \text{MGD})\left(10^6 \ \frac{\text{gal}}{\text{MG}}\right)}{2000 \ \frac{\text{gal}}{\text{day-ft}^2}}$$
$$= 7000 \ \text{ft}^2$$

Maximum daily flow controls. The diameter based on maximum daily flow is

$$D = \sqrt{\frac{4A}{\pi}} = \sqrt{\frac{(4)(8750 \text{ ft}^2)}{\pi}}$$
$$= 105.5 \text{ ft} \quad (110 \text{ ft})$$

Check weir loading. The weir length is

$$L = \pi D = \pi(110 \text{ ft})$$
$$= 345.6 \text{ ft}$$

The weir loading at peak hourly flow is

$$q_{\text{weir}} = \frac{Q_{\text{max}}}{L} = \frac{(14 \text{ MGD})\left(10^6 \frac{\text{gal}}{\text{MG}}\right)}{345.6 \text{ ft}}$$
$$= 40{,}509 \text{ gal/day-ft}$$

This result for weir loading would not work, because it is greater than the 35,000 gal/day-ft criterion. The weir length needed is

$$L_{\text{min}} = \frac{Q_{\text{max}}}{q_{\text{weir}}} = \frac{(14 \text{ MGD})\left(10^6 \frac{\text{gal}}{\text{MG}}\right)}{35{,}000 \frac{\text{gal}}{\text{day-ft}}}$$
$$= 400 \text{ ft}$$

The required clarifier diameter based on weir length is

$$D = \frac{L}{\pi} = \frac{400 \text{ ft}}{\pi}$$
$$= 127.3 \text{ ft}$$

Check the detention time with a 130 ft diameter. The problem gives the sidewall depth, h. The tank volume is

$$V = \frac{\pi D^2 h}{4} = \frac{\pi(130 \text{ ft})^2(8 \text{ ft})}{4}$$
$$= 106{,}186 \text{ ft}^3$$

The detention time is

$$t_d = \frac{V}{Q}$$
$$= \frac{(106{,}186 \text{ ft}^3)\left(24 \frac{\text{hr}}{\text{day}}\right)\left(60 \frac{\text{min}}{\text{hr}}\right)\left(7.48 \frac{\text{gal}}{\text{ft}^3}\right)}{(14 \text{ MGD})\left(10^6 \frac{\text{gal}}{\text{MG}}\right)}$$
$$= 81 \text{ min}$$

This detention time is greater than the given minimum detention time of 60 min, so this result is okay. The clarifier diameter that meets all criteria is 130 ft.

The answer is (D).

43. The MLSS in the sludge waste line is

$$\text{MLSS} = \frac{\left(1000 \frac{\text{mg}}{\text{g}}\right)\left(1000 \frac{\text{mL}}{\text{L}}\right)}{\text{SVI}}$$
$$= \frac{\left(1000 \frac{\text{mg}}{\text{g}}\right)\left(1000 \frac{\text{mL}}{\text{L}}\right)}{100 \frac{\text{mL}}{\text{g}}}$$
$$= 10\,000 \text{ mg/L}$$

The MLVSS in the sludge waste line is

$$X_w = PC$$

P is the percent volatile solids. C is the MLSS concentration.

$$X_w = \left(0.70 \frac{\text{MLVSS}}{\text{MLSS}}\right)\left(10\,000 \frac{\text{mg}}{\text{L}}\right)$$
$$= 7000 \text{ mg/L MLVSS}$$

V_a is the volume of aeration in m^3, X is the mixed liquor VSS in mg/L, Q_e is the effluent flow rate in L/s, X_e is the effluent VSS in mg/L, and Q_w is the wasting flow rate in L/s. The sludge age, or mean cell residence time, θ_c, is

$$\theta_c = \frac{V_a X}{Q_e X_e + Q_w X_w}$$

$$= \frac{(4300 \text{ m}^3)\left(1000 \frac{\text{L}}{\text{m}^3}\right)\left(2500 \frac{\text{mg}}{\text{L}} \text{ MLSS}\right) \times \left(0.70 \frac{\text{MLVSS}}{\text{MLSS}}\right)}{\left(200 \frac{\text{L}}{\text{s}}\right)\left(3600 \frac{\text{s}}{\text{h}}\right)\left(10 \frac{\text{mg}}{\text{L}}\right)\left(0.70 \frac{\text{MLVSS}}{\text{MLSS}}\right) + \left(20 \frac{\text{L}}{\text{s}}\right)\left(3600 \frac{\text{s}}{\text{h}}\right)\left(7000 \frac{\text{mg}}{\text{L}}\right)}$$
$$= 14.8 \text{ h} \quad (15 \text{ h})$$

The answer is (C).

44. The basin volume based on detention time is

$$V = Qt$$
$$= \left(2 \frac{\text{MG}}{\text{day}}\right)\left(10^6 \frac{\text{gal}}{\text{MG}}\right)\left(\frac{1 \text{ ft}^3}{7.48 \text{ gal}}\right)$$
$$\times (90 \text{ min})\left(\frac{1 \text{ day}}{24 \text{ hr}}\right)\left(\frac{1 \text{ hr}}{60 \text{ min}}\right)$$
$$= 16{,}711 \text{ ft}^3$$

The area based on detention time is

$$A = \frac{V}{h} = \frac{16{,}711 \text{ ft}^3}{8 \text{ ft}}$$
$$= 2089 \text{ ft}^2$$

The diameter based on detention time is

$$D = \sqrt{\frac{4A}{\pi}} = \sqrt{\frac{(4)\,(2089 \text{ ft}^2)}{\pi}}$$
$$= 51.6 \text{ ft}$$

The area based on overflow rate is

$$A = \frac{Q}{\text{v}^*} = \frac{(2 \text{ MGD})\left(10^6 \dfrac{\text{gal}}{\text{MG}}\right)}{600 \dfrac{\text{gal}}{\text{day-ft}^2}}$$
$$= 3333 \text{ ft}^2$$

The diameter based on v^* is

$$D = \sqrt{\frac{4A}{\pi}} = \sqrt{\frac{(4)(3333 \text{ ft}^2)}{\pi}}$$
$$= 65.1 \text{ ft} \quad [\text{controls}]$$

Check the weir loading for a 65 ft diameter. The weir length is

$$L = \pi D = \pi(65 \text{ ft})$$
$$= 204.2 \text{ ft}$$

The weir loading is

$$q_{\text{weir}} = \frac{Q}{L} = \frac{(2 \text{ MGD})\left(10^6 \dfrac{\text{gal}}{\text{MG}}\right)}{204.2 \text{ ft}}$$
$$= 9794 \text{ gal/day-ft}$$

The weir loading is acceptable. The diameter of the basin is 65 ft.

The answer is (C).

45. The mass of available alum is

$$m_{\text{alum}} = C_{\text{alum}} P$$

C_{alum} is the concentration of alum, and P is the fraction of available alum.

$$m_{\text{alum}} = \left(1.400 \,\frac{\text{kg}}{\text{L}}\right)(0.49)$$
$$= 0.686 \text{ kg/L}$$

The atomic weight of aluminum is 26.98 g/mol. The aluminum available is

$$C_{\text{Al}} = m_{\text{alum}}(\text{WR})$$

WR is the ratio of mass of aluminum to mass of available alum.

$$C_{\text{Al}} = \left(0.686 \,\frac{\text{kg alum}}{\text{L solution}}\right)\left(2 \,\frac{\text{mol Al}}{\text{mol alum}}\right)$$
$$\times \left(26.98 \,\frac{\text{g Al}}{\text{mol Al}}\right)\left(\frac{1 \text{ mol alum}}{666.7 \text{ g alum}}\right)$$
$$\times \left(1000 \,\frac{\text{g}}{\text{kg}}\right)$$
$$= 55.5 \text{ kg Al/L solution}$$

The reaction for phosphate precipitation with aluminum is

$$\text{Al}^{+3} + \text{H}_n\text{PO}_4^{3-n} \rightleftharpoons \text{AlPO}_4 + n\text{H}^+$$

One mole of aluminum is required to precipitate 1 mol of phosphorus.

The atomic weight of phosphorus is 30.97 g/mol. The theoretical unit alum dose required is

$$D_{\text{alum}} = \frac{m_{\text{Al}}}{m_{\text{P}}}$$
$$= \left(\frac{1 \text{ mol Al}}{1 \text{ mol P}}\right)\left(\frac{1 \text{ mol P}}{30.97 \text{ g P}}\right)\left(26.98 \,\frac{\text{g Al}}{\text{mol Al}}\right)$$
$$= 0.871 \text{ g Al/g P}$$

R is the ratio of actual to theoretical dose required. The unit volume of alum required is

$$V_{\text{unit}} = \frac{D_{\text{alum}} R}{C_{\text{Al}}}$$
$$= \frac{\left(0.871 \,\dfrac{\text{g Al}}{\text{g P}}\right)\left(1.50 \,\dfrac{\text{actual dose}}{\text{theoretical dose}}\right)}{55.5 \,\dfrac{\text{g Al}}{\text{L alum solution}}}$$
$$= 0.0235 \text{ L alum solution/g P}$$

The volume of alum solution required is

$$V_{\text{total}} = QCV_{\text{unit}}$$
$$= \left(400 \,\frac{\text{L}}{\text{s}}\right)\left(86\,400 \,\frac{\text{s}}{\text{d}}\right)\left(10 \,\frac{\text{mg P}}{\text{L}}\right)$$
$$\times \left(\frac{1 \text{ g}}{1000 \text{ mg}}\right)\left(0.0235 \,\frac{\text{L alum solution}}{\text{g P}}\right)$$
$$= 8122 \text{ L/d alum solution}$$

The answer is (B).

46. The overall mass of TSS removed is

$$m_{\text{TSS}} = (C_i - C_e)Q$$

C_i is the influent TSS concentration in mg/L, C_e is the effluent TSS concentration in mg/L, and Q is the flow in L/s.

$$m_{\text{TSS}} = \left(220 \ \frac{\text{mg}}{\text{L}} - 20 \ \frac{\text{mg}}{\text{L}}\right)\left(400 \ \frac{\text{L}}{\text{s}}\right)\left(86\,400 \ \frac{\text{s}}{\text{d}}\right)$$
$$\times \left(\frac{1 \ \text{kg}}{10^6 \ \text{mg}}\right)$$
$$= 6912 \ \text{kg/d}$$

The BOD entering the aeration process is

$$C_a = C_i F$$

F is the fraction of BOD entering aeration process.

$$C_a = \left(280 \ \frac{\text{mg}}{\text{L}}\right)\left(\frac{100\% - 30\%}{100\%}\right)$$
$$= 196 \ \text{mg/L}$$

The mass of biological solids produced is

$$m_{\text{bio}} = Y(\Delta\text{BOD})Q$$

Y is the cell yield in kg/kg.

$$m_{\text{bio}} = \left(\frac{60 \ \text{kg suspended solids removed}}{100 \ \text{kg BOD removed}}\right)$$
$$\times \left(196 \ \frac{\text{mg}}{\text{L}} \ \text{BOD} - 20 \ \frac{\text{mg}}{\text{L}} \ \text{BOD}\right)\left(400 \ \frac{\text{L}}{\text{s}}\right)$$
$$\times \left(86\,400 \ \frac{\text{s}}{\text{d}}\right)\left(\frac{1 \ \text{kg}}{10^6 \ \text{mg}}\right)$$
$$= 3650 \ \text{kg suspended biological solids}$$

The total dry sludge mass produced is

$$m_{\text{total}} = m_{\text{TSS}} + m_{\text{bio}}$$
$$= 6912 \ \frac{\text{kg}}{\text{d}} + 3650 \ \frac{\text{kg}}{\text{d}}$$
$$= 10\,562 \ \text{kg/d} \quad (11\,000 \ \text{kg/d})$$

The answer is (D).

47. The mass of sludge wasted is

$$P_x = Q_{\text{waste}} C_{\text{waste}}$$

Q_{waste} is the waste flow rate. C_{waste} is the waste suspended solids concentration.

$$P_x = \left(30 \ \frac{\text{m}^3}{\text{d}}\right)\left(1000 \ \frac{\text{L}}{\text{m}^3}\right)\left(10\,000 \ \frac{\text{mg}}{\text{L}}\right)\left(\frac{1 \ \text{kg}}{10^6 \ \text{mg}}\right)$$
$$= 300 \ \text{kg/d}$$

The volume of methane produced is

$$V = \left(0.35 \ \frac{\text{m}^3 \ \text{CH}_4}{\text{kg}}\right)(EQS_o - 1.42P_x)$$
$$= \left(0.35 \ \frac{\text{m}^3 \ \text{CH}_4}{\text{kg}}\right)$$
$$\times \left(\left(\frac{(0.7)\left(3000 \ \frac{\text{m}^3}{\text{d}}\right)\left(400 \ \frac{\text{mg}}{\text{L}}\right)}{\times \left(\frac{1 \ \text{g}}{1000 \ \text{mg}}\right)\left(1000 \ \frac{\text{L}}{\text{m}^3}\right)}{1000 \ \frac{\text{g}}{\text{kg}}} \right) - (1.42)\left(300 \ \frac{\text{kg}}{\text{d}}\right) \right)$$
$$= 144.9 \ \text{m}^3/\text{d} \quad (140 \ \text{m}^3/\text{d})$$

The answer is (A).

48. The reactions are

$$\text{Cl}_2 + \text{H}_2\text{O} \rightleftharpoons \text{HOCl} + \text{H}^+ + \text{Cl}^-$$
$$\text{SO}_2 + \text{HOCl} + \text{H}_2\text{O} \rightleftharpoons \text{Cl}^- + \text{SO}_4^{-2} + 3\text{H}^+$$

The dose of SO_2 is

$$D_{\text{SO}_2} = \left(6 \ \frac{\text{mg}}{\text{L}} \ \text{Cl}_2\right)\left(\frac{1 \ \text{mol Cl}_2}{71 \ \text{g Cl}_2}\right)\left(1 \ \frac{\text{mol HOCl}}{\text{mol Cl}_2}\right)$$
$$\times \left(52.5 \ \frac{\text{g HOCl}}{\text{mol HOCl}}\right)\left(1 \ \frac{\text{mol SO}_2}{\text{mol HOCl}}\right)$$
$$\times \left(\frac{1 \ \text{mol HOCl}}{52.5 \ \text{g HOCl}}\right)\left(64 \ \frac{\text{g SO}_4}{\text{mol SO}_4}\right)$$
$$= 5.41 \ \text{mg/L}$$

The mass per day of SO_2 is

$$\dot{m} = CQ$$
$$= \left(5.41 \ \frac{\text{mg}}{\text{L}}\right)\left(80 \ \frac{\text{L}}{\text{s}}\right)\left(86\,400 \ \frac{\text{s}}{\text{d}}\right)\left(\frac{1 \ \text{kg}}{10^6 \ \text{mg}}\right)$$
$$= 37.4 \ \text{kg/d} \quad (40 \ \text{kg/d})$$

The answer is (D).

49. The overall denitrification rate is

$$U'_{dn} = U_{dn}1.09^{T-20}(1 - DO)$$

U_{dn} is the specific denitrification rate in (kg NO$_3$-N)/(kg·d MLVSS), T is the wastewater temperature in °C, and DO is the wastewater dissolved oxygen in mg/L.

$$U'_{dn} = \left(0.09 \; \frac{\text{kg NO}_3\text{-N}}{\text{kg·d MLVSS}}\right)(1.09^{10-20})$$
$$\times \left(1 \; \frac{\text{mg}}{\text{L}} - 0.2 \; \frac{\text{mg}}{\text{L}}\right)$$
$$= 0.030 \text{ kg NO}_3\text{-N/kg·d MLVSS}$$

The detention time is

$$\theta = \frac{S_o - S}{U'_{dn}X}$$

S_o is the influent substrate concentration in mg/L, S is the effluent substrate concentration in mg/L, and X is the MLVSS in mg/L.

$$\theta = \frac{\left(26 \; \frac{\text{mg}}{\text{L}} - 4 \; \frac{\text{mg}}{\text{L}}\right)\left(24 \; \frac{\text{h}}{\text{d}}\right)}{\left(0.030 \; \frac{\text{kg NO}_3\text{-N}}{\text{kg·d MLVSS}}\right)\left(2500 \; \frac{\text{mg}}{\text{L}}\right)}$$
$$= 7.04 \text{ h} \quad (7 \text{ h})$$

The answer is (C).

50. The area of each port is

$$A = \frac{\pi D^2}{4} = \frac{\pi (12 \text{ in})^2 \left(\frac{1 \text{ ft}}{12 \text{ in}}\right)^2}{4}$$
$$= 0.785 \text{ ft}^2$$

The velocity from each port is

$$v_d = \frac{Q}{A(\text{number of ports})}$$
$$= \frac{75 \; \frac{\text{ft}^3}{\text{sec}}}{\left(0.785 \text{ ft}^2\right)(10 \text{ ports})}$$
$$= 9.55 \text{ ft/sec}$$

The initial dilution is

$$S = \left(\frac{UHL}{2Q_d}\right)\left(1 + \sqrt{1 + \frac{2Q_d v_d \cos\theta}{U^2 LH}}\right)$$

S is the initial dilution, U is the river velocity in ft/sec, H is the river depth in ft, L is the diffuser length in ft, Q_d is the discharge flow rate, and θ is the orientation of ports above horizontal.

$$20 = \left(\frac{\left(3 \; \frac{\text{ft}}{\text{sec}}\right)(4 \text{ ft} + 2 \text{ ft})L}{(2)\left(75 \; \frac{\text{ft}^3}{\text{sec}}\right)}\right)$$
$$\times \left(1 + \sqrt{1 + \frac{(2)\left(75 \; \frac{\text{ft}^3}{\text{sec}}\right)\left(9.55 \; \frac{\text{ft}}{\text{sec}}\right)\cos 60°}{\left(3 \; \frac{\text{ft}}{\text{sec}}\right)^2 L(4 \text{ ft} + 2 \text{ ft})}}\right)$$

Solve by iteration or calculator solver.

$$L = 80 \text{ ft}$$

The answer is (D).

51. The hydraulic detention time is

$$t = \frac{LWnd}{Q}$$

L is the basin length, W is the basin width, n is the fraction of the cross-sectional area not occupied by plants, d is the basin depth, and Q is the average flow rate in ft^3/day.

$$t = \frac{(200 \text{ ft})(2500 \text{ ft})(0.75)(2 \text{ ft})}{40,000 \; \frac{\text{ft}^3}{\text{day}}}$$
$$= 18.75 \text{ days}$$

The effluent BOD is

$$C_e = C_o A e^{-0.7K_T A_v^{1.75} t}$$

C_o is the influent BOD$_5$, A is the coefficient that represents the fraction of BOD$_5$ not removed by settling at the head of the system, K_T is the temperature-dependent, first-order rate constant in d^{-1}, and A_v is the specific surface area for microbiological activity in ft^2/ft^3.

$$K_T = K_{20}(1.1)^{T-20}$$
$$= (0.006 \text{ d}^{-1})(1.1)^{10-20}$$
$$= 0.0023 \text{ d}^{-1}$$

$$C_e = \left(200 \; \frac{\text{mg}}{\text{L}}\right)(0.52)e^{\left(\begin{array}{c}(-0.7)(0.0023 \text{ d}^{-1})\\ \times\left(4.8 \text{ ft}^2/\text{ft}^3\right)^{1.75}(18.75 \text{ d})\end{array}\right)}$$
$$= 64.5 \text{ mg/L} \quad (65 \text{ mg/L})$$

The answer is (B).

52. The air-to-solids ratio is

$$\frac{A}{S} = \frac{\left(1.3 \dfrac{\text{mg air}}{\text{mL air}}\right) a_s (f p_{\text{abs}} - 1)}{S_s}$$

a_s is the air solubility, f is the fraction of air dissolved at a given pressure, p_{abs} is the absolute pressure in atmospheres, and S_s is the suspended solids concentration.

The absolute pressure is

$$p_{\text{abs}} = \frac{\dfrac{S_s \dfrac{A}{S}}{\left(1.3 \dfrac{\text{mg air}}{\text{mL air}}\right) a_s} + 1}{f}$$

$$= \frac{\dfrac{\left(300 \dfrac{\text{mg}}{\text{L}}\right)\left(0.07 \dfrac{\text{mg air}}{\text{mg solids}}\right)}{\left(1.3 \dfrac{\text{mg air}}{\text{mL air}}\right)\left(18.7 \dfrac{\text{mL air}}{\text{L water}}\right)} + 1}{0.6}$$

$$= 3.11 \text{ atm}$$

The gage pressure is

$$p_{\text{gage}} = p_{\text{abs}} - p_{\text{atm}}$$

p_{atm} is the pressure of one atmosphere.

$$p_{\text{gage}} = (3.11 \text{ atm})\left(\frac{101 \text{ kPa}}{1 \text{ atm}}\right) - 101 \text{ kPa}$$

$$= 213.1 \text{ kPa} \quad (220 \text{ kPa})$$

The answer is (D).

53. The statements in options (A), (B), and (C) are true. They describe conditions that can cause high concentrations of suspended solids to accumulate in wastewater effluent.

The statement in option (D) is false. Excessive return rates can cause dilute return-activated sludge that will add to the hydraulic load through aeration. This condition reduces removal efficiency in sedimentation tanks and, thus, is detrimental to the process. However, it is not normally a cause of high levels of suspended solids in effluent.

The answer is (D).

54. The BOD of the dilution in bottle no. 1 is

$$\text{BOD} = \frac{\text{initial DO} - \text{final DO}}{\text{fraction dilution}}$$

$$= \frac{8.6 \dfrac{\text{mg}}{\text{L}} - 6.0 \dfrac{\text{mg}}{\text{L}}}{\dfrac{1.00\%}{100\%}}$$

$$= 260 \text{ mg/L}$$

The BOD levels of the remaining dilutions are given in the following table.

bottle no.	waste-water portion (mg)	dilution (%)	initial DO (mg/L)	final DO (mg/L)	BOD (mg/L)	group average BOD (mg/L)
1	3	1	8.6	6.0	260	
2	3	1	8.6	6.1	250	
3	3	1	8.4	6.6	180	230
4	2	0.67	8.5	6.1	358	
5	2	0.67	8.5	6.4	313	
6	2	0.67	8.6	6.0	388	353
7	1	0.33	8.6	6.8	545	
8	1	0.33	8.5	6.4	636	
9	1	0.33	8.4	6.5	575	585

Plot the average BOD versus dilution. The data show increasing BOD with increasing dilution, which indicates the presence of toxicity in the wastewater. Biological activity is inhibited at the lower dilutions (1.0% wastewater) compared to higher dilutions (0.33% wastewater).

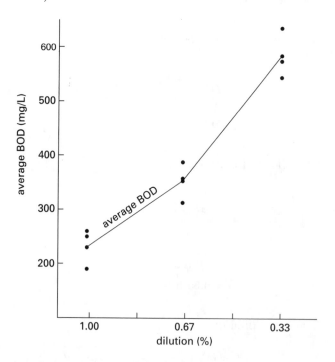

The answer is (D).

55. Statement II is false. Algae produce oxygen. This aids oxidation ponds by producing part of the oxygen required by bacteria.

Statement III is false. Phosphorus, as well as nitrogen and carbon, must be controlled to meet the water quality objectives of limiting algae growth.

Statements I, IV, and V are true.

The answer is (B).

56. The correct matches are as follows.

I. Aquatic life diversity and abundance can be limited by the absence of sunlight or the presence of pollution (2), and they require the presence of oxygen, carbon dioxide, nitrogen, and phosphorus (5).

II. Algae and green plants are primary producers, since they use the energy of sunlight to synthesize inorganic substances into living tissue (7).

III. Fly nymphs, copepods, and water fleas are first-order consumers in the food chain (1).

IV. Sunfish are second-order consumers of small herbivores (3).

V. Bass, pike, and salmon are third-order consumers of flesh-eaters (6).

VI. Decomposition and decay of fish and other aquatic life release nutrients that are recycled by algae through photosynthesis (4).

The answer is (A).

57. The stream velocity is

$$v = \frac{Q}{A} = \frac{\left(400\ \frac{L}{s}\right)\left(\frac{1\ m^3}{1000\ L}\right)}{(1\ m)(10\ m)}$$
$$= 0.04\ m/s$$

The travel time is

$$t = \frac{L}{v}$$
$$= \frac{(10\ km)\left(1000\ \frac{m}{km}\right)\left(\frac{1\ min}{60\ s}\right)\left(\frac{1\ h}{60\ min}\right)}{0.04\ \frac{m}{s}}$$
$$= 69.4\ h$$

The decimal fraction of bacteria remaining is

$$\frac{N}{N_o} = (1 + nkt)^{-1/n}$$

N is the modal number of bacteria remaining, N_o is the modal number of intial bacteria, n is the coefficient of nonuniformity or retardation for the specific watercourse, and k is the intial die-away for the specific bacteria population in the receiving water.

$$\frac{N}{N_o} = \left(1 + (6.15)\left(\frac{1500}{h}\right)(69.4\ h)\right)^{-1/6.15}$$
$$= 0.11$$

The percent removal is

$$P = \left(1 - \frac{N}{N_o}\right) \times 100\%$$
$$= (1 - 0.11) \times 100\%$$
$$= 89\%\quad(90\%)$$

The answer is (D).

58. Statement I is false. Nitrogen in the form of ammonia from conventional wastewater treatment plants adds organic loading that increases bacterial decomposition.

Statement II is false. Disinfection only temporarily depletes microbiological populations, which subsequently rapidly increase and exert oxygen demand relative to the organic load.

Statements III, IV, and V are true.

The answer is (D).

59. The BOD_5 of the stream after complete mixing with the effluent is

$$C_f = \frac{C_1 Q_1 + C_2 Q_2}{Q_1 + Q_2}$$

C_f is the final mixed concentration, C_1 is the effluent concentration, C_2 is the stream concentration before mixing, Q_1 is the effluent flow, and Q_2 is the stream flow before mixing.

$$BOD_{5,20°C} = \frac{\left(60\ \frac{mg}{L}\right)\left(0.5\ \frac{m^3}{s}\right) + \left(4.5\ \frac{mg}{L}\right)\left(6\ \frac{m^3}{s}\right)}{0.5\ \frac{m^3}{s} + 6\ \frac{m^3}{s}}$$
$$= 8.77\ mg/L$$

Use the same equation twice more to find the DO and the temperature of the stream after complete mixing with the effluent. For the variables C_f, C_1, and C_2, substitute the appropriate values of DO and temperature, respectively. The DO of the stream after complete mixing with the effluent is

$$DO_{mix} = \frac{\left(2\ \frac{mg}{L}\right)\left(0.5\ \frac{m^3}{s}\right) + \left(8.5\ \frac{mg}{L}\right)\left(6\ \frac{m^3}{s}\right)}{0.5\ \frac{m^3}{s} + 6\ \frac{m^3}{s}}$$

$$= 8.0\ mg/L$$

The temperature of the stream after complete mixing is

$$T = \frac{(25°C)\left(0.5\ \frac{m^3}{s}\right) + (15°C)\left(6\ \frac{m^3}{s}\right)}{0.5\ \frac{m^3}{s} + 6\ \frac{m^3}{s}}$$

$$= 15.77°C$$

The deoxygenation rate constant after complete mixing is

$$K_{d,T} = K_{d,20°C}(1.046)^{T-20}$$
$$= \left(0.15\ d^{-1}\right)(1.046)^{15.77-20}$$
$$= 0.124\ d^{-1}$$

The reaeration rate constant after complete mixing is

$$K_{r,T} = K_{r,20°C}(1.024)^{T-20}$$
$$= \left(0.25\ d^{-1}\right)(1.024)^{15.77-20}$$
$$= 0.226\ d^{-1}$$

The ultimate BOD of the stream after complete mixing at 20°C is

$$BOD_{u,20°C} = \frac{BOD_{5,20°C}}{1 - 10^{-K_d t}}$$
$$= \frac{8.77\ \frac{mg}{L}}{1 - 10^{-(0.15\ d^{-1})(5\ d)}}$$
$$= 10.67\ mg/L$$

The ultimate BOD of the stream after complete mixing at 15.77°C is

$$BOD_{u,T} = BOD_{u,20°C}(0.02T + 0.6)$$
$$= \left(10.67\ \frac{mg}{L}\right)((0.02)(15.77°C) + 0.6)$$
$$= 9.77\ mg/L$$

The oxygen deficit after complete mixing is

$$D_O = DO_{sat} - DO_{mix}$$

DO_{sat} is the dissolved oxygen saturation, and DO_{mix} is the dissolved oxygen after complete mixing.

$$D_O = 9.95\ \frac{mg}{L} - 8.0\ \frac{mg}{L}$$
$$= 1.95\ mg/L$$

The time to the critical point is

$$t_c = \left(\frac{1}{K_r - K_d}\right)$$
$$\times \log_{10}\left(\left(\frac{K_d BOD_u - K_r D_O + K_d D_O}{K_d BOD_u}\right)\left(\frac{K_r}{K_d}\right)\right)$$
$$= \left(\frac{1}{0.226\ d^{-1} - 0.124\ d^{-1}}\right)$$
$$\times \log_{10}\left(\begin{pmatrix}\dfrac{\left(0.124\ d^{-1}\right)\left(9.77\ \frac{mg}{L}\right) \\ - \left(0.226\ d^{-1}\right)\left(1.95\ \frac{mg}{L}\right) \\ + \left(0.124\ d^{-1}\right)\left(1.95\ \frac{mg}{L}\right)}{\left(0.124\ d^{-1}\right)\left(9.77\ \frac{mg}{L}\right)} \\ \times \left(\dfrac{0.226\ d^{-1}}{0.124\ d^{-1}}\right)\end{pmatrix}\right)$$

$$= 1.792\ d$$

The critical oxygen deficit is

$$D_c = \left(\frac{K_d BOD_u}{K_r}\right)10^{-K_d t_c}$$
$$= \left(\frac{\left(0.124\ d^{-1}\right)\left(9.77\ \frac{mg}{L}\right)}{0.226\ d^{-1}}\right)$$
$$\times \left(10^{-\left(0.124\ d^{-1}\right)(1.79\ d)}\right)$$
$$= 3.22\ mg/L$$

The DO at the critical point is

$$DO_c = DO_{sat} - D_c$$
$$= 9.95\ \frac{mg}{L} - 3.22\ \frac{mg}{L}$$
$$= 6.73\ mg/L \quad (6.8\ mg/L)$$

The answer is (C).

60. Statements I, III, and IV are true. As statement III suggests, algae are able to fix nitrogen from the air, so phosphorus levels have been identified as more critical to controlling eutrophication.

Statement II is false. It often takes a long time to reverse or retard the effects of eutrophication because the long water-renewal time (detention time) reduces flushing action.

Statement V is false. Decaying vegetation or algae can accumulate in bottom sediments where decomposition depletes DO and affects aquatic life.

The answer is (B).

61. The rate constant at 30°C is

$$K_t = K_{20°C}\theta^{T-20}$$

K_t is the rate constant at a specified temperature in degrees Celsius, $K_{20°C}$ is the rate constant at 20°C, and θ is the temperature constant, which is typically 1.047 for a temperature between 20°C and 30°C.

$$K_t = \left(0.23 \text{ d}^{-1}\right)(1.047)^{30-20}$$
$$= 0.364 \text{ d}^{-1}$$

The 5 d BOD is

$$\text{BOD}_5 = L\left(1 - e^{-K_t t}\right)$$
$$= \left(150 \frac{\text{mg}}{\text{L}}\right)\left(1 - e^{(-0.364 \text{ d}^{-1})(5 \text{ d})}\right)$$
$$= 125.7 \text{ mg/L} \quad (125 \text{ mg/L})$$

The answer is (B).

62. Statements I, II, and V are true.

Statement III is false. The membrane filter technique results in a count of the number of colonies of coliform bacteria. A confirmation test is not required.

Statement IV is false. *Aerobacter aerogenes* is a practical indicator organism even though it is commonly found in soil.

The answer is (B).

63. The detention time is

$$t = \frac{V}{Q} = \frac{1440 \text{ m}^3}{\left(400 \frac{\text{L}}{\text{s}}\right)\left(\frac{1 \text{ m}^3}{1000 \text{ L}}\right)\left(60 \frac{\text{s}}{\text{min}}\right)}$$
$$= 60 \text{ min}$$

The required CT value, $C_t t$, can be found with the following equation.

$$\frac{N_t}{N_o} = (1 + 0.23C_t t)^{-3}$$

N_t is the number of coliform organisms after treatment at time t, N_o is the number of coliform organisms before treatment, and C_t is the total amperometric chlorine residual at time t in mg/L.

$$\frac{\dfrac{200}{100 \text{ mL}}}{\dfrac{2 \times 10^8}{100 \text{ mL}}} = (1 + 0.23C_t t)^{-3}$$
$$C_t t = 430.4 \text{ mg/L·min}$$

For a time of 60 min, C_t is

$$C_t = \frac{430.4 \dfrac{\text{mg}}{\text{L·min}}}{60 \text{ min}}$$
$$= 7.17 \text{ mg/L} \quad (7.2 \text{ mg/L})$$

The answer is (D).

64. Statement I is false. This is an effluent toxicity test species.

Statement IV is false. This is an effluent toxicity test species.

Statement VII is false. This is a bacteria known for oxidizing the taste and odor compound geosmin caused by blue-green algae.

Statements II, III, V, and VI are true.

The answer is (B).

65. When more than three dilutions are employed in a decimal series of dilutions, the results from only three of these are used in computing the MPN. The three dilutions selected are the highest dilution (i.e., the lowest sample portion) giving positive results in all five portions tested (no lower dilution with any negative results), and the two next succeeding higher dilutions. The dilution corresponding to the middle dilution is used to calculate the MPN from the MPN index. The three dilutions used are given in the following table.

serial dilution	sample portion (mL)	number of positive reactions
2	0.01	5
3	0.001	2
4	0.0001	1

The middle dilution is 0.001 mL. Using the MPN tables, the MPN index is 70 MPN/100 mL.

The MPN is

$$\text{MPN} = \frac{I}{D}$$

I is the MPN index in MPN/100 mL. D is the middle dilution corresponding to 1 mL of a series of 10 mL, 1 mL, and 0.1 mL.

$$\text{MPN} = \frac{\dfrac{70 \text{ MPN}}{100 \text{ mL}}}{0.001 \dfrac{\text{mL}}{\text{mL}}}$$
$$= 70\,000/100 \text{ mL}$$

The answer is (D).

66. The decimal fraction of the wastewater sample used for dilution no. 1 is

$$P = \frac{S}{T}$$

S is the sample volume. T is the total volume of the bottle.

$$P = \frac{5 \text{ mL}}{300 \text{ mL}}$$
$$= 0.0167$$

The decimal fraction of the wastewater sample used for dilution no. 2 is

$$P = \frac{10 \text{ mL}}{300 \text{ mL}}$$
$$= 0.0333$$

The decimal fraction of the wastewater sample used for dilution no. 3 is

$$P = \frac{15 \text{ mL}}{300 \text{ mL}}$$
$$= 0.050$$

The 5 d BOD is

$$\text{BOD}_5 = \frac{D_1 - D_2}{P}$$

D_1 is the initial DO in mg/L, and D_2 is the final DO after 5 d in mg/L. The 5 d BOD of dilution no. 1 is

$$\text{BOD}_5 = \frac{8.0 \dfrac{\text{mg}}{\text{L}} - 6.2 \dfrac{\text{mg}}{\text{L}}}{0.0167}$$
$$= 107.8 \text{ mg/L}$$

The 5 d BOD of dilution no. 2 is

$$\text{BOD}_5 = \frac{8.2 \dfrac{\text{mg}}{\text{L}} - 5.2 \dfrac{\text{mg}}{\text{L}}}{0.0333}$$
$$= 90.0 \text{ mg/L}$$

The 5 d BOD of dilution no. 3 is

$$\text{BOD}_5 = \frac{8.4 \dfrac{\text{mg}}{\text{L}} - 3.5 \dfrac{\text{mg}}{\text{L}}}{0.050}$$
$$= 98.0 \text{ mg/L}$$

The average 5 d BOD of the three dilutions is

$$\text{BOD}_{\text{avg}} = \frac{\sum \text{BOD}}{n}$$

BOD_n represents individual BOD results. n is the number of samples.

$$\text{BOD}_{\text{avg}} = \frac{107.8 \dfrac{\text{mg}}{\text{L}} + 90.0 \dfrac{\text{mg}}{\text{L}} + 98.0 \dfrac{\text{mg}}{\text{L}}}{3}$$
$$= 98.6 \text{ mg/L}$$

The ultimate BOD is

$$\text{BOD}_{\text{ult}} = \frac{\text{BOD}_5}{1 - e^{-k_1 t}}$$

k_1 is the rate constant to base e, and t is the test duration in d.

$$\text{BOD}_{\text{ult}} = \frac{98.6 \dfrac{\text{mg}}{\text{L}}}{1 - e^{-(0.25 \text{ d}^{-1})(5 \text{ d})}}$$
$$= 138 \text{ mg/L} \quad (140 \text{ mg/L})$$

The answer is (C).

67. Statements I and V are true.

Statement II is false. Although copper sulfate can kill algae that are in high density in a reservoir, this practice does not lead to improved biological quality.

Statement III is false. Changing the hydrography to reduce stream velocity would typically decrease stream reaeration and have a negative effect on biological life.

Statement IV is false. Chlorination of natural waters would adversely affect stream biology.

The answer is (B).

68. The total mass of the weekly solid waste generation per household is

$$m_{\text{week-hh}} = P_{\text{hh}} m_{\text{person-day}}$$

P_{hh} is the persons per household. $m_{\text{person-day}}$ is the mass of solid waste per person per day.

$$m_{\text{week-hh}} = \left(3.5 \, \frac{\text{persons}}{\text{household}}\right)\left(7 \, \frac{\text{days}}{\text{wk}}\right)\left(4 \, \frac{\text{lbm}}{\text{person-day}}\right)$$
$$= 98 \, \text{lbm/wk-household}$$

The total weekly mass of separated paper is

$$m_{\text{week}} = m_{\text{week-hh}} F_{\text{paper/sw}} N_{\text{hh}} R_{\text{paper}}$$

$F_{\text{paper/sw}}$ is the mass fraction of paper to total solid waste, N_{hh} is the number of households, and R_{paper} is the rate of participation of paper recycling.

$$m_{\text{week}} = \left(98 \, \frac{\text{lbm total solid waste}}{\text{wk·household}}\right)$$
$$\times \left(\frac{21 \, \text{lbm paper}}{100 \, \text{lbm total solid waste}}\right)$$
$$\times (2000 \, \text{households}) \left(\frac{80\%}{100\%} \, \text{participation}\right)$$
$$= 32{,}928 \, \text{lbm paper/wk}$$

The total weekly volume of paper is

$$V_{\text{week}} = \frac{m_{\text{week}}}{\rho_{\text{paper}}}$$

ρ_{paper} is the density of separated paper.

$$V_{\text{week}} = \frac{32{,}928 \, \dfrac{\text{lbm paper}}{\text{wk}}}{6 \, \dfrac{\text{lbm}}{\text{ft}^3}}$$
$$= 5488 \, \text{ft}^3/\text{wk paper}$$

The volumes for the other recylables are given in the table.

recyclable	volume (ft^3/wk)
paper	5488
cardboard	2091
plastic	490
glass	245
tin cans	196
total	8510

The volume of recyclables transported each trip is

$$V_{\text{trip}} = C_{\text{trip}} E_{\text{trip}}$$

C_{trip} is the vehicle capacity per trip, and E_{trip} is the efficiency of vehicle utilization.

$$V_{\text{trip}} = \left(15 \, \frac{\text{yd}^3}{\text{trip}}\right)\left(\frac{90\%}{100\%}\right)\left(27 \, \frac{\text{ft}^3}{\text{yd}^3}\right)$$
$$= 364.5 \, \text{ft}^3/\text{trip}$$

The number of trips is

$$N_{\text{trip}} = \frac{V_{\text{wk-total}}}{V_{\text{trip}}}$$

$V_{\text{wk-total}}$ is the total volume to be transported per week.

$$N_{\text{trip}} = \frac{8510 \, \dfrac{\text{ft}^3}{\text{wk}}}{364.5 \, \dfrac{\text{ft}^3}{\text{trip}}}$$
$$= 23.3 \, \text{trips} \quad (24 \, \text{trips})$$

The answer is (C).

69. The average number of packer trucks is

$$N_{\text{packer-day}} = m_{\text{day}} P_{\text{packer}} L_{\text{packer}}$$

m_{day} is the mass of waste handled through transfer station daily, P_{packer} is the percent of station wastes handled by packer trucks, and L_{packer} is the loading capacity of packer trucks.

$$N_{\text{packer-day}} = \left(500 \, \frac{\text{U.S. tons}}{\text{day}}\right)\left(\frac{80\%}{100\%}\right)\left(\frac{1 \, \text{packer}}{6 \, \text{U.S. tons}}\right)$$
$$= 66.6 \, \text{packer/day}$$

The average number of small vehicles is

$$N_{\text{small veh-day}} = m_{\text{day}} P_{\text{small veh}} L_{\text{small veh}}$$

$P_{\text{small veh}}$ is the percent of station wastes handled by small vehicles, and $L_{\text{small veh}}$ is the loading capacity of small vehicles.

$$N_{\text{small veh-day}} = \left(500 \, \frac{\text{U.S. tons}}{\text{day}}\right)\left(\frac{20\%}{100\%}\right)$$
$$\times \left(\frac{1 \, \text{small veh}}{0.4 \, \text{U.S. tons}}\right)$$
$$= 250 \, \text{small veh/day}$$

The peak hourly number of packer trucks is

$$N_{\text{packer-hr}} = N_{\text{packer-day}} F_{\text{peaking-mo}} F_{\text{peaking-hr}}$$

$F_{\text{peaking-mo}}$ is the monthly peaking factor, and $F_{\text{peaking-hr}}$ is the hourly peaking factor.

$$N_{\text{packer-hr}} = \left(66.6 \ \frac{\text{average mo packer}}{\text{day}}\right)\left(\frac{1 \text{ day}}{8 \text{ hr}}\right)$$
$$\times \left(1.5 \ \frac{\text{peak mo}}{\text{average mo}}\right)$$
$$\times \left(2.0 \ \frac{\text{peak hr}}{\text{average hr}}\right)$$
$$= 25 \text{ packer trucks/hr}$$

The peak hourly number of small vehicles is

$$N_{\text{small veh-hr}} = N_{\text{small veh-day}} F_{\text{peaking-mo}} F_{\text{peaking-hr}}$$
$$= \left(250 \ \frac{\text{average mo small veh}}{\text{day}}\right)\left(\frac{1 \text{ day}}{8 \text{ hr}}\right)$$
$$\times \left(1.0 \ \frac{\text{peak mo}}{\text{average mo}}\right)$$
$$\times \left(1.0 \ \frac{\text{peak hr}}{\text{average hr}}\right)$$
$$= 31.2 \text{ small vehicles/hr}$$

The number of bays for packer trucks is

$$N_{\text{packer-bays}} = N_{\text{packer-hr}} U_{\text{packer}}$$

$N_{\text{packer-hr}}$ is the number of packers per hour, and U_{packer} is the unloading time per packer per bay.

$$N_{\text{packer-bays}} = \left(25 \ \frac{\text{packers}}{\text{hr}}\right)\left(6 \ \frac{\text{min-bay}}{\text{packer}}\right)\left(\frac{1 \text{ hr}}{60 \text{ min}}\right)$$
$$= 2.5 \text{ bays} \quad (3 \text{ bays})$$

The number of bays for small vehicles is

$$N_{\text{small veh-bays}} = N_{\text{small veh-hr}} U_{\text{small veh}}$$

$N_{\text{small veh-hr}}$ is the number of small vehicles per hour, and $U_{\text{small veh}}$ is the unloading time per small vehicle.

$$N_{\text{small veh-bays}} = \left(31.2 \ \frac{\text{small veh}}{\text{hr}}\right)\left(15 \ \frac{\text{min}}{\text{small veh}}\right)$$
$$\times \left(\frac{1 \text{ hr}}{60 \text{ min}}\right)$$
$$= 7.8 \text{ bays} \quad (8 \text{ bays})$$

The total number of bays is

$$3 \text{ bays} + 8 \text{ bays} = 11 \text{ bays}$$

The answer is (B).

70. The number of moles of material initially present is

$$n_{\text{initial}} = \frac{m_{\text{sw-initial}}}{\text{MW}}$$

$m_{\text{sw-initial}}$ is the initial mass of solid waste, and MW is the molecular weight in kg/mol.

$$n_{\text{initial}} = \frac{1000 \text{ kg}}{\left(\begin{array}{c}(40)\left(12 \ \frac{\text{kg C}}{\text{mol}}\right) + (55)\left(1 \ \frac{\text{kg H}}{\text{mol}}\right) \\ + (30)\left(16 \ \frac{\text{kg O}}{\text{mol}}\right)\end{array}\right)}$$
$$= 0.985 \text{ mol solid waste}$$

The number of moles of residue is

$$n_{\text{final}} = \frac{m_{\text{sw-final}}}{\text{MW}}$$

$m_{\text{sw-final}}$ is the final mass of solid waste.

$$n_{\text{final}} = \frac{350 \text{ kg}}{\left(\begin{array}{c}(12)\left(12 \ \frac{\text{kg C}}{\text{mol}}\right) + (25)\left(1 \ \frac{\text{kg H}}{\text{mol}}\right) \\ + (10)\left(16 \ \frac{\text{kg O}}{\text{mol}}\right)\end{array}\right)}$$
$$= 1.064 \text{ mol residue}$$

The number of moles of material leaving the process per mole of material entering is

$$n = \frac{n_{\text{sw-final}}}{n_{\text{sw-initial}}}$$
$$= \frac{1.064 \text{ mol}}{0.985 \text{ mol}}$$
$$= 1.08$$

The aerobic stabilization process is

$$C_a H_b O_c N_d + 0.5\,(ny + 2s + r - c)\,O_2$$
$$\rightarrow nC_w H_x O_y N_z + sCO_2 + rH_2O + (d - nz)\,NH_3$$
$$r = 0.5\,(b - nx - 3(d - nz))$$
$$s = a - nw$$

The values for a, b, c, and d for the initial material $C_{40}H_{55}O_{30}$ are

$$a = 40$$
$$b = 55$$
$$c = 30$$
$$d = 0$$

The values for w, x, y, and z for the residual $C_{12}H_{25}O_{10}$ are

$$w = 12$$
$$x = 25$$
$$y = 10$$
$$z = 0$$

The value of r is

$$r = 0.5\,(b - nx - 3(d - nz))$$
$$= (0.5)\,(55 - (1.08)(25) - (3)(0 - (1.08)(0)))$$
$$= 14$$

The value of s is

$$s = a - nw$$
$$= 40 - (1.08)(12)$$
$$= 27$$

The mass of oxygen required is

$$m_{O_2} = 0.5\,(ny + 2s + r - c)\,(\text{moles } O_2)(\text{MW}_{O_2})$$
$$= (0.5)\,((1.08)(10) + (2)(27) + 14 - 30)$$
$$\times (0.985 \text{ mol } O_2)\left(32\,\frac{\text{kg}}{\text{mol}}\right)$$
$$= 769 \text{ kg } O_2 \quad (770 \text{ kg } O_2)$$

The answer is (D).

71. The mass of solid waste per square yard is

$$\frac{m_{sw}}{A} = \rho_{\text{lift}} P_{sw} d_{\text{lift}}$$

ρ_{lift} is the density of lift, P_{sw} is the proportion of solid waste, d_{lift} is the depth of lift, and A is the area.

$$\frac{m_{sw}}{A} = \left(1000\,\frac{\text{lbm}}{\text{yd}^3}\right)\left(\frac{5 \text{ parts solid waste}}{6 \text{ parts total}}\right)$$
$$\times (12 \text{ ft})\left(\frac{1 \text{ yd}}{3 \text{ ft}}\right)$$
$$= 3333 \text{ lbm/yd}^2$$

The moisture in the solid waste per square yard is

$$\frac{w_{sw}}{A} = \frac{m_{sw}}{A} P_{\text{moist}}$$

w_{sw} is the mass of solid waste per square yard, and P_{moist} is the percent moisture.

$$\frac{w_{sw}}{A} = \left(3333\,\frac{\text{lbm}}{\text{yd}^2}\right)\left(\frac{20\%}{100\%}\right)$$
$$= 667 \text{ lbm water/yd}^2$$

The mass of dry solid waste per square yard is

$$\frac{m_{\text{dry}}}{A} = \frac{m_{sw}}{A} P_{\text{dry}}$$

P_{dry} is the percent dry material.

$$\frac{m_{\text{dry}}}{A} = \left(3333\,\frac{\text{lbm}}{\text{yd}^2}\right)\left(\frac{80\%}{100\%}\right)$$
$$= 2667 \text{ lbm dry solid waste/yd}^2$$

The mass of soil per square yard is

$$\frac{m_{\text{soil}}}{A} = \rho_{\text{lift}} P_{\text{soil}} d_{\text{lift}}$$

P_{soil} is the proportion of soil.

$$\frac{m_{\text{soil}}}{A} = \left(3000\,\frac{\text{lbm}}{\text{yd}^3}\right)\left(\frac{1 \text{ parts soil}}{6 \text{ parts total}}\right)$$
$$\times (12 \text{ ft})\left(\frac{1 \text{ yd}}{3 \text{ ft}}\right)$$
$$= 2000 \text{ lbm/yd}^2$$

The average mass of the lift is

$$m_{\text{lift}} = \frac{d_{sw}}{d_{\text{lift}}} m_{sw} + \frac{d_{\text{soil}}}{d_{\text{lift}}} m_{\text{soil}}$$

d_{sw} is the depth of the solid waste, and d_{soil} is the depth of the soil.

$$m_{\text{lift}} = \left(\frac{10 \text{ ft}}{12 \text{ ft}}\right)\left(3333\,\frac{\text{lbm}}{\text{yd}^2}\right) + \left(\frac{2 \text{ ft}}{12 \text{ ft}}\right)\left(2000\,\frac{\text{lbm}}{\text{yd}^2}\right)$$
$$= 3111 \text{ lbm/yd}^2$$

The field capacity is

$$\text{FC} = 0.6 - 0.55\left(\frac{m_{\text{lift}}}{10{,}000 + m_{\text{lift}}}\right)$$
$$= 0.6 - (0.55)\left(\frac{3111\,\dfrac{\text{lbm}}{\text{yd}^2}}{10{,}000 + 3111\,\dfrac{\text{lbm}}{\text{yd}^2}}\right)$$
$$= 0.469$$

The water than can be held in the lift is

$$\text{FC}_{sw} = \text{FC}(\text{dry mass of solid waste})$$
$$= (0.469)\left(2667\,\frac{\text{lbm}}{\text{yd}^2}\right)$$
$$= 1251 \text{ lbm/yd}^2$$

The leachate formed is

$$\text{leachate} = w_{sw} - FC_{sw}$$
$$= 667 \ \frac{\text{lbm}}{\text{yd}^2} - 1251 \ \frac{\text{lbm}}{\text{yd}^2}$$
$$= -584 \ \text{lbm/yd}^2$$

Because the field capacity of the solid waste is greater than the actual moisture, no leachate will form.

The answer is (A).

72. The mass of paper to be recycled per 100 kg of solid waste is

$$m_{\text{paper recycled}} = m_{\text{paper collected}} E_{\text{paper}}$$

$m_{\text{paper collected}}$ is the mass of paper as collected, and E_{paper} is the effectiveness of the paper recycling program.

$$m_{\text{paper recycled}} = (35 \ \text{kg}) \left(\frac{70\%}{100\%} \right)$$
$$= 24.5 \ \text{kg}$$

The cardboard to be recycled per 100 kg of solid waste is

$$m_{\text{cardboard recycled}} = (m_{\text{cardboard collected}})(E_{\text{cardboard}})$$
$$= (5 \ \text{kg}) \left(\frac{80\%}{100\%} \right)$$
$$= 4 \ \text{kg}$$

The composition of the waste stream before and after paper and cardboard recycling is given in the following table.

component	mass as collected with no recycling (kg)	mass of recycled materials (kg)	mass as collected after recycling (kg)	components collected after recycling (%)
food waste	10	0	10	14.0
paper	35	24.5	10.5	14.7
cardboard	5	4	1.0	1.3
plastics	8	0	8	11.2
textiles	2	0	2	2.8
yard waste	20	0	20	28.0
glass	8	0	8	11.2
all other	12	0	12	16.8
total	100	28.5	71.5	100.0

The yard waste will be 28% of the waste stream after the recycling of paper and cardboard.

The answer is (B).

73. The direct haul cost per unit volume and unit time is

$$C'_{\text{dh}} = \frac{C_{\text{dh}}}{L_{\text{dh}}}$$

C_{dh} is the cost of direct haul per minute, and L_{dh} is the loading capacity of direct haul.

$$C'_{\text{dh}} = \left(\frac{\$40}{1 \ \text{hr}} \right) \left(\frac{1 \ \text{hr}}{60 \ \text{min}} \right) \left(\frac{1 \ \text{direct haul vehicle}}{15 \ \text{yd}^3} \right)$$
$$= \$0.044/\text{yd}^3 \cdot \text{min}$$

The semi-trailer haul cost per unit volume and unit time is

$$C'_{\text{st}} = \frac{C_{\text{st}}}{L_{\text{st}}}$$

C_{st} is the cost of semi-trailer haul per minute, and L_{st} is the loading capacity of semi-trailer.

$$C'_{\text{st}} = \left(\frac{\$50}{1 \ \text{hr}} \right) \left(\frac{1 \ \text{hr}}{60 \ \text{min}} \right) \left(\frac{1 \ \text{semi-trailer vehicle}}{100 \ \text{yd}^3} \right)$$
$$= \$0.0083/\text{yd}^3 \cdot \text{min}$$

Plot the round trip time versus the cost per cubic yard for each system. C'_{dh} and C'_{st} are the line slopes. Start the semi-trailer plot at the transfer station operation cost.

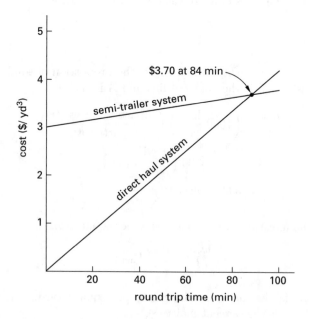

The direct haul cost is equivalent to the cost of the semi-trailer system at a round trip time of 84 min (85 min).

The answer is (A).

74. The gross energy available from the solid waste is

$$E_{\text{gross}} = m_{\text{sw}} E_{\text{sw}}$$

m_{sw} is the mass of solid waste, and E_{sw} is the energy content of solid waste.

$$E_{\text{gross}} = \frac{\left(1500\ \dfrac{\text{U.S. tons}}{\text{day}}\right)\left(2000\ \dfrac{\text{lbm}}{\text{U.S. ton}}\right) \times \left(4800\ \dfrac{\text{Btu}}{\text{lbm}}\right)}{24\ \dfrac{\text{hr}}{\text{day}}}$$

$$= 600 \times 10^6\ \text{Btu/hr}$$

The steam energy available from the boiler is

$$E_{\text{steam}} = E_{\text{gross}} \eta_{\text{steam}}$$

η_{steam} is the steam efficiency of boiler.

$$E_{\text{steam}} = \left(600 \times 10^6\ \frac{\text{Btu}}{\text{hr}}\right)\left(\frac{72\%}{100\%}\right)$$

$$= 432 \times 10^6\ \text{Btu/hr}$$

The mechanical energy available from the turbine is

$$E_{\text{turbine}} = E_{\text{steam}} \eta_{\text{turbine}}$$

η_{turbine} is the mechanical efficiency of turbine.

$$E_{\text{turbine}} = \left(432 \times 10^6\ \frac{\text{Btu}}{\text{hr}}\right)\left(\frac{28\%}{100\%}\right)$$

$$= 121 \times 10^6\ \text{Btu/hr}$$

The gross electric power generation is

$$P_{\text{gross}} = E_{\text{turbine}} \eta_{\text{electrical}}$$

$\eta_{\text{electrical}}$ is the electrical generation efficiency.

$$P_{\text{gross}} = \frac{\left(121 \times 10^6\ \dfrac{\text{Btu}}{\text{hr}}\right)\left(\dfrac{92\%}{100\%}\right)}{3413\ \dfrac{\text{Btu}}{\text{kW·hr}}}$$

$$= 32\,614\ \text{kW}$$

The station service allowance is

$$P_{\text{station}} = (32\,614\ \text{kW})\left(\frac{4\%}{100\%}\right)$$

$$= 1305\ \text{kW}$$

The unaccounted for heat loss is

$$P_{\text{loss}} = (32\,614\ \text{kW})\left(\frac{6\%}{100\%}\right)$$

$$= 1957\ \text{kW}$$

The net electric power available is

$$P_{\text{net}} = P_{\text{gross}} - P_{\text{station}} - P_{\text{loss}}$$

$$= 32\,614\ \text{kW} - 1305\ \text{kW} - 1957\ \text{kW}$$

$$= 29\,352\ \text{kW}$$

The overall efficiency is

$$\eta_{\text{overall}} = \frac{P_{\text{net}}}{E_{\text{gross}}}$$

$$= \frac{(29\,352\ \text{kW})\left(3413\ \dfrac{\text{Btu}}{\text{kW·hr}}\right) \times 100\%}{600 \times 10^6\ \dfrac{\text{Btu}}{\text{hr}}}$$

$$= 16.7\% \quad (17\%)$$

The answer is (C).

75. The frequency of exposure is

$$\text{FOE} = \frac{t_{\text{exposure}}}{(1\ \text{yr})\left(365\ \dfrac{\text{d}}{\text{yr}}\right)}$$

t_{exposure} is the exposure time measured in days.

$$\text{FOE} = \frac{120\ \text{d}}{365\ \text{d}}$$

$$= 0.329$$

The absorbed dose, AB, is 1.0.

The drinking water equivalent level in mg/L is

$$\text{DWEL} = \frac{D_{\text{ref}}(\text{ABW})}{(\text{DWI})(\text{AB})(\text{FOE})}$$

D_{ref} is the oral reference dose for MTBE in mg/kg·d, ABW is the average body mass in kg, and DWI is the daily water intake in L.

$$\text{DWEL} = \frac{\left(0.005\ \dfrac{\text{mg}}{\text{kg·d}}\right)(70\ \text{kg})}{\left(2\ \dfrac{\text{L}}{\text{d}}\right)(1.0)(0.333)}$$

$$= 0.53\ \text{mg/L}$$

The answer is (C).

76. Statement I is false. RCRA Subtitle C addresses hazardous waste and RCRA Subtitle D addresses non-hazardous wastes.

Statement III is false. USEPA drinking water standards for MCLs apply to groundwater protection from sanitary landfills at any location.

Statements II, IV, and V are true.

The answer is (D).

77. The flow per meter of gallery is

$$q = \frac{0.5K(H^2 - h^2)}{L}$$

K is the coefficient of permeability in m/s, H is the water surface height at distance L in meters, h is the water surface height at gallery in meters, and L is the distance to water surface H in meters.

$$Q = \frac{(0.5)\left(0.15 \frac{\text{cm}}{\text{s}}\right)\left(\frac{1 \text{ m}}{100 \text{ cm}}\right)\left((6 \text{ m})^2 - (0.5 \text{ m})^2\right)}{10 \text{ m}}$$
$$= 2.68 \times 10^{-3} \text{ m}^3/\text{m·s} \quad (3 \times 10^{-3} \text{ m}^3/\text{m·s})$$

The answer is (D).

78. The thickness of the saturated aquifer at the radius of influence is

$$y_1 = y_{\text{inf}} - y_{\text{bottom}}$$

y_{inf} is the phreatic surface elevation at the radius of influence, and y_{bottom} is the elevation of the bottom of the aquifer.

$$y_1 = 100 \text{ m} - 55 \text{ m} = 45 \text{ m}$$

The thickness of the saturated aquifer at the well is

$$y_2 = y_{\text{inf}} - y_{\text{bottom}} - y_{\text{drawdown}}$$

y_{drawdown} is the drawdown in the well.

$$y_2 = 100 \text{ m} - 55 \text{ m} - 3.5 \text{ m} = 41.5 \text{ m}$$

The radius of the well is

$$r_2 = \frac{D_{\text{well}}}{2}$$

D_{well} is the diameter of the well.

$$r_2 = \left(\frac{450 \text{ mm}}{2}\right)\left(\frac{1 \text{ m}}{1000 \text{ mm}}\right)$$
$$= 0.225 \text{ m}$$

Using Depuit's equation, the hydraulic conductivity is

$$K = \frac{Q \ln \frac{r_1}{r_2}}{\pi (y_1^2 - y_2^2)}$$
$$= \frac{\left(50 \frac{\text{L}}{\text{s}}\right)\left(\frac{1 \text{ m}^3}{1000 \text{ L}}\right)\ln\left(\frac{300 \text{ m}}{0.225 \text{ m}}\right)}{\pi \left((45 \text{ m})^2 - (41.5 \text{ m})^2\right)}$$
$$= 378.26 \times 10^{-6} \text{ m/s} \quad (400 \times 10^{-6} \text{ m/s})$$

The answer is (B).

79. Total hardness includes primarily calcium and magnesium ions, and to a lesser extent (because of normally lower concentrations) iron, manganese, strontium, and aluminum. The concentrations of hardness-causing ions are

$$Ca^{2+} = 90 \text{ mg/L}$$
$$Mg^{2+} = 38 \text{ mg/L}$$
$$Fe^{2+} = 0.10 \text{ mg/L}$$

$$E = \frac{M}{Z}$$

E is the equivalent weight, and M is the molecular weight. Z is the absolute value of the ion change: the number of H^+ or OH^- ions a species can react with or yield in an acid-base reaction, or the absolute value of the change in valence occccurring in an oxidation-reduction reaction.

The millequivalents are

$$Ca^{2+} = \frac{40 \text{ mg}}{2} = 20 \text{ mg/meq}$$
$$Mg^{2+} = \frac{24 \text{ mg}}{2} = 12 \text{ mg/meq}$$
$$Fe^{2+} = \frac{56 \text{ mg}}{2} = 28 \text{ mg/meq}$$

The hardness ions in meq/L are

$$H = \frac{C}{E}$$

H is the hardness in meq/L, C is the concentration in mg/L, and E is the millequivalents of hardness-causing ions.

$$Ca^{2+} = \frac{90 \frac{\text{mg}}{\text{L}}}{20 \frac{\text{mg}}{\text{meq}}} = 4.50 \text{ meq/L}$$

$$Mg^{2+} = \frac{38 \frac{\text{mg}}{\text{L}}}{12 \frac{\text{mg}}{\text{meq}}} = 3.17 \text{ meq/L}$$

$$Fe^{2+} = \frac{0.10 \frac{\text{mg}}{\text{L}}}{28 \frac{\text{mg}}{\text{meq}}} = 0.0036 \text{ meq/L}$$

The millequivalent weight of $CaCO_3$ is

$$E_{CaCO_3} = \frac{40 \text{ mg} + 12 \text{ mg} + (3)(16 \text{ mg})}{2}$$
$$= 50 \text{ mg/meq}$$

The total hardness as $CaCO_3$ is

$$\left(4.50 \ \frac{\text{meq}}{\text{L}} + 3.17 \ \frac{\text{meq}}{\text{L}} + 0.0036 \ \frac{\text{meq}}{\text{L}} \right)$$
$$\times \left(50 \ \frac{\text{mg CaCO}_3}{\text{meq}} \right)$$
$$= 384 \text{ mg/L as } CaCO_3 \quad (380 \text{ mg/L as } CaCO_3)$$

The answer is (A).

80. The soil permeability is

$$K = \left(350 \ \frac{\text{m}}{\text{yr}} \right) \left(\frac{1 \text{ yr}}{365 \text{ d}} \right) \left(\frac{1 \text{ d}}{86 \, 400 \text{ s}} \right)$$
$$= 1.1 \times 10^{-5} \text{ m/s}$$

The total flow for each drain is given by the rational equation for radial flow. L is the length of the drain in meters, H is the height of the water table above the invert of drain at the midpoint of drain spacing in meters, h is the height of the water level above the drain invert in the pipe in meters, S is the spacing between drains in meters, and d is the diameter of the drains in meters.

$$Q = \frac{\pi K L (H - h)}{2.3 \log \dfrac{S}{D}}$$
$$= \frac{\pi \left(1.1 \times 10^{-5} \ \dfrac{\text{m}}{\text{s}} \right) (1200 \text{ m}) (3.5 \text{ m} - 0.3 \text{ m})}{2.3 \log \left(\dfrac{100 \text{ m}}{0.3 \text{ m}} \right)}$$
$$= 0.022 \, 87 \text{ m}^3/\text{s} \quad (0.025 \text{ m}^3/\text{s})$$

The answer is (C).

Solutions
Geotechnical

81. Consolidation is calculated at point A, the midpoint of the second clay stratum (layer 2). It is reasonable to assume that the increase in pressure at the midpoint of the layer can be computed using the 2:1 stress distribution method, and that the stress spread starts at the lower third point of the piles.

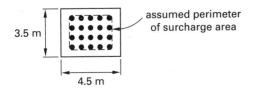

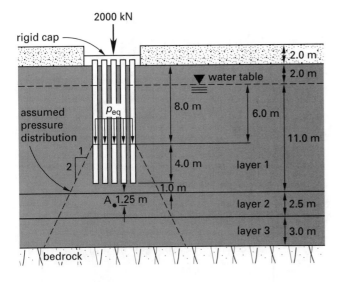

The equivalent pressure at the lower third of the piles, p_{eq}, is

$$p_{eq} = \frac{P}{BL}$$
$$= \frac{2000 \text{ kN}}{(3.5 \text{ m})(4.5 \text{ m})}$$
$$= 127.0 \text{ kN/m}^2$$

To determine the initial pressure, p_o, calculate the pressure from each soil layer individually at the points of interest.

The initial pressure at point A is

$$p_{oA} = \sum \gamma_i z_i$$
$$= \left(17.5 \frac{\text{kN}}{\text{m}^3}\right)(2 \text{ m}) + \left(19.0 \frac{\text{kN}}{\text{m}^3}\right)(2 \text{ m})$$
$$+ \left(19.0 \frac{\text{kN}}{\text{m}^3} - 9.81 \frac{\text{kN}}{\text{m}^3}\right)(11.0 \text{ m})$$
$$+ \left(18.0 \frac{\text{kN}}{\text{m}^3} - 9.81 \frac{\text{kN}}{\text{m}^3}\right)(1.25 \text{ m})$$
$$= 184.3 \text{ kN/m}^2 \quad (184.3 \text{ kPa})$$

To determine the change in pressures, Δp, calculate the pressure change at the midpoint of layer 2 using the 2:1 stress distribution method, and assume that the stress spread starts at the lower third point of the piles. The change in pressure at point A is

$$\Delta p_A = \left(127.0 \frac{\text{kN}}{\text{m}^2}\right) \left(\frac{(3.5 \text{ m})(4.5 \text{ m})}{(4.5 \text{ m} + (2)(6.25 \text{ m})(0.50))} \times \left(\begin{array}{c} 3.5 \text{ m} + (2) \\ \times (6.25 \text{ m})(0.50) \end{array} \right) \right)$$
$$= 19.1 \text{ kN/m}^2 \quad (19.1 \text{ kPa})$$

For a layer of soil with thickness H, the settlement S is

$$S = \frac{C_c}{1 + e_0} H \log_{10} \frac{p_0 + \Delta p}{p_0}$$

The settlement in clay layer 2 is

$$S_2 = \left(\frac{0.32}{1 + 1.03}\right)(2.5 \text{ m})$$
$$\times \log_{10}\left(\frac{184.3 \frac{\text{kN}}{\text{m}^2} + 19.1 \frac{\text{kN}}{\text{m}^2}}{184.3 \frac{\text{kN}}{\text{m}^2}}\right)$$
$$= 0.017 \text{ m} \quad (17 \text{ mm})$$

The answer is (B).

82. The bentonite content is based on dry weight. The dry unit weight is

$$\gamma_d = \frac{\gamma}{1+w} = \frac{17.0 \, \frac{kN}{m^3}}{1+0.18}$$
$$= 14.4 \, kN/m^3$$

Each lift has a thickness, t, of 125 mm. The total layer weight, W_t, of each lift is

$$W_{total} = t\gamma_d$$
$$= (125 \, mm)\left(\frac{1 \, m}{1000 \, mm}\right)\left(14.4 \, \frac{kN}{m^3}\right)$$
$$= 1.8 \, kN/m^2$$

As 8% (by weight) of this mixture is bentonite, the total weight of bentonite required per lift is

$$W_{bentonite} = 0.08W_{total}$$
$$= (0.08)\left(1.8 \, \frac{kN}{m^2}\right)$$
$$= 0.144 \, kN/m^2 \quad (0.14 \, kN/m^2)$$

The answer is (A).

83. The hydraulic gradient, i, can be calculated from the leachate head, h, and the thickness of the clay liner, t.

$$i = \frac{H}{t} = \frac{0.5 \, m + 1.2 \, m}{1.2 \, m}$$
$$= 1.42 \, m/m$$

The effective velocity is given by

$$v = Ki = \left(6 \times 10^{-7} \, \frac{cm}{s}\right)\left(\frac{1 \, m}{100 \, cm}\right)\left(1.42 \, \frac{m}{m}\right)$$
$$= 8.52 \times 10^{-9} \, m/s$$

The flow for the entire landfill on an annual basis is

$$Q = vA$$
$$= \left(8.52 \times 10^{-9} \, \frac{m}{s}\right)\left(3.16 \times 10^7 \, \frac{s}{yr}\right)$$
$$\times (300 \, m)(400 \, m)$$
$$= 3.2 \times 10^4 \, m^3/yr \quad (10^4 \, m^3/yr)$$

The answer is (B).

84. Find the unit weights of the soils. The dry unit weight of sand is

$$\gamma_d = \frac{\gamma}{1+w} = \frac{18.4 \, \frac{kN}{m^3}}{1+0.18}$$
$$= 15.6 \, kN/m^3$$

The unit weight of sand above the water table (after lowering the water table) is

$$\gamma = \gamma_d(1+w) = \left(15.6 \, \frac{kN}{m^3}\right)(1+0.11)$$
$$= 17.3 \, kN/m^3$$

The dry unit weight of clay is

$$\gamma_d = \frac{SG\gamma_w}{1+e} = \frac{(2.68)\left(9.81 \, \frac{kN}{m^3}\right)}{1+1.15}$$
$$= 12.2 \, kN/m^3$$

The water content of clay is

$$w = \frac{Se}{SG} = \frac{(1.0)(1.15)}{2.68}$$
$$= 0.43$$

The saturated unit weight of clay is

$$\gamma_{sat} = \gamma_d(1+w) = \left(12.2 \, \frac{kN}{m^3}\right)(1+0.43)$$
$$= 17.4 \, kN/m^3$$

The initial effective stress at the midpoint of the clay layer is

$$\bar{\sigma}_o = \sigma - u$$
$$= \left(\begin{array}{c} \left(18.4 \, \frac{kN}{m^3} - 9.81 \, \frac{kN}{m^3}\right)(30.0 \, m) \\ + \left(17.4 \, \frac{kN}{m^3} - 9.81 \, \frac{kN}{m^3}\right)(4.0 \, m) \end{array} \right)$$
$$= 288 \, kPa$$

The final effective stress at the midpoint of the clay layer is

$$\bar{\sigma}_f = \sigma - u$$
$$= \left(17.3 \, \frac{kN}{m^3}\right)(16.3 \, m) + \left(18.4 \, \frac{kN}{m^3} - 9.81 \, \frac{kN}{m^3}\right)$$
$$\times (13.7 \, m) + \left(17.4 \, \frac{kN}{m^3} - 9.81 \, \frac{kN}{m^3}\right)(4.0 \, m)$$
$$= 430 \, kN/m^2 \quad (430 \, kPa)$$

The ultimate settlement of the clay layer is

$$\Delta H = \frac{H}{1+e} C_c \log \frac{\overline{\sigma}_f}{\overline{\sigma}_o}$$
$$= \left(\frac{8.0 \text{ m}}{1+1.15} \right) (0.32) \left(\log \frac{430 \text{ kPa}}{288 \text{ kPa}} \right)$$
$$= 0.21 \text{ m} \quad (0.20 \text{ m})$$

The answer is (B).

85. For a falling-head permeameter, the coefficient of permeability is

$$K = \frac{A'l}{At} \ln \frac{h_i}{h_f}$$

The ratios of the areas can be found from the ratio of the diameters. A is the cross-sectional area of the soil, A' is the cross-sectional area of the standpipe.

$$\frac{A'}{A} = \frac{\pi \left(\frac{d'}{2} \right)^2}{\pi \left(\frac{d}{2} \right)^2} = \left(\frac{d'}{d} \right)^2$$
$$= \left(\frac{0.25 \text{ cm}}{10.0 \text{ cm}} \right)^2$$
$$= 0.000\,625$$
$$K = \left(\frac{A'}{A} \right) \left(\frac{l}{t} \right) \log \frac{h_i}{h_f}$$
$$= (0.000\,625) \left(\frac{6 \text{ cm}}{1278 \text{ s}} \right) \left(\ln \frac{100 \text{ cm}}{50 \text{ cm}} \right)$$
$$= 2 \times 10^{-6} \text{ cm/s}$$

The answer is (D).

86. The time required to achieve 90% consolidation, t_{90}, is

$$t_{90} = \frac{T_{90} H^2}{C_v}$$

This can be rearranged to give

$$T_{90} = \frac{C_v t_{90}}{H^2}$$

For the clayey silt sample,

$$t_{90,s} = (10 \text{ min}) \left(\frac{60 \text{ s}}{1 \text{ min}} \right) + 46 \text{ s}$$
$$= 646 \text{ s}$$
$$H_s = (5 \text{ cm}) \left(\frac{1 \text{ m}}{100 \text{ cm}} \right)$$
$$= 0.05 \text{ m}$$
$$T_{90} = \frac{C_v t_{90,s}}{H_s^2}$$
$$= \frac{C_v (646 \text{ s})}{(0.05 \text{ m})^2}$$

For the 25 m clayey silt layer,

$$T_{90} = \frac{C_v t_{90,l}}{H_l^2}$$
$$= \frac{C_v t_{90,l}}{(25 \text{ m})^2}$$

Since T_{90} is the same for both the site soil and the sample soil, the right hand sides of the last two equations are also equivalent

$$\frac{C_v (646 \text{ s})}{(0.05 \text{ m})^2} = \frac{C_v t_{90,t}}{(25 \text{ m})^2}$$
$$t_{90,l} = \frac{(646 \text{ s})(25 \text{ m})^2}{(0.05 \text{ m})^2} \left(\frac{1 \text{ min}}{60 \text{ s}} \right) \left(\frac{1 \text{ h}}{60 \text{ min}} \right)$$
$$\times \left(\frac{1 \text{ d}}{24 \text{ h}} \right) \left(\frac{1 \text{ y}}{365 \text{ d}} \right)$$
$$= 5.12 \text{ yr} \quad (5 \text{ yr})$$

The answer is (C).

87. For the settlement of 124 mm, the degree of consolidation is

$$U_z = \frac{\Delta H}{\Delta H_{\text{ult}}} = \frac{124 \text{ mm}}{502 \text{ mm}}$$
$$= 0.247$$

The time factor, T_v, can be found in a table of approximate time factors, or calculated from the following equation.

$$T_v = \tfrac{1}{4} \pi U_z^2 \quad [U_z < 0.60]$$
$$= \tfrac{1}{4} \pi (0.247)^2$$
$$= 0.048$$

The time for a layer to reach a specific consolidation is

$$t = \frac{T_v H^2}{C_v}$$
$$5 \text{ yr} = \frac{0.048 H^2}{C_v}$$
$$\frac{H^2}{C_v} = 104 \text{ yr}$$

For the total settlement of 250 mm,

$$U_z = \frac{\Delta H}{\Delta H_{\text{ult}}} = \frac{250 \text{ mm}}{502 \text{ mm}}$$
$$= 0.498$$

The time factor is

$$T_v = \tfrac{1}{4} \pi (0.498)^2$$
$$= 0.195$$

The time to reach the settlement of 250 mm is

$$t = T_v \frac{H^2}{C_v}$$
$$= (0.195)(104 \text{ yr})$$
$$= 20.3 \text{ yr}$$

The remaining time to reach a settlement of 250 mm is

$$\Delta t = 20.3 \text{ yr} - 5 \text{ yr}$$
$$= 15.3 \text{ yr} \quad (15 \text{ yr})$$

The answer is (C).

88. Since the backfill is horizontal and the retaining wall is smooth, the coefficient of active earth pressure is

$$k_a = \tan^2 \left(45° - \frac{\phi}{2} \right)$$
$$= \tan^2 \left(45° - \frac{32°}{2} \right)$$
$$= 0.31$$

To determine the total resultant force and the location of the moment arm, divide the backfill into five components.

For the surcharge at the backfill surface, the effective stress is

$$\sigma'_{a,A} = k_a \sigma'_v = (0.31)(25 \text{ kPa})$$
$$= 7.75 \text{ kPa}$$

For the surcharge and sand above the water table,

$$\sigma'_{a,B} = \sigma'_{a,A} + k_a \gamma H$$
$$= 7.75 \text{ kPa} + (0.31) \left(19.5 \frac{\text{kN}}{\text{m}^3} \right) (3.5 \text{ m})$$
$$= 28.9 \text{ kPa}$$

For the surcharge, sand above the water table, and sand below the water table,

$$\sigma'_{a,C} = \sigma'_{a,B} + k_a (\gamma_{\text{sat}} - \gamma_w) H$$
$$= 28.9 \text{ kPa} + (0.31) \left(20.3 \frac{\text{kN}}{\text{m}^3} - 9.81 \frac{\text{kN}}{\text{m}^3} \right)$$
$$\quad \times (6.5 \text{ m})$$
$$= 50.0 \text{ kPa}$$

For the surcharge, sand above the water table, sand below the water table, and water pore pressure,

$$\sigma'_{a,C} + u = \sigma'_{a,C} + \gamma_w H$$
$$= 50.0 \text{ kPa} + \left(9.81 \frac{\text{kN}}{\text{m}^3} \right) (6.5 \text{ m})$$
$$= 113.8 \text{ kPa}$$

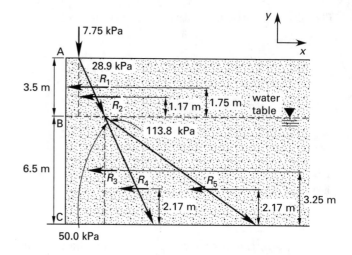

The resultant forces are the areas of the geometrically-shaped pressure distributions.

$$R_1 = (7.75 \text{ kPa})(3.5 \text{ m}) = 27.1 \text{ kN/m}$$
$$R_2 = \left(\frac{1}{2} \right) (28.9 \text{ kPa} - 7.75 \text{ kPa})(3.5 \text{ m})$$
$$= 37.0 \text{ kN/m}$$
$$R_3 = (28.9 \text{ kPa})(6.5 \text{ m}) = 187.9 \text{ kN/m}$$
$$R_4 = \left(\frac{1}{2} \right) (50.0 \text{ kPa} - 28.9 \text{ kPa})(6.5 \text{ m})$$
$$= 68.6 \text{ kN/m}$$
$$R_5 = \left(\frac{1}{2} \right) (113.8 \text{ kPa} - 50.0 \text{ kPa})(6.5 \text{ m})$$
$$= 207.4 \text{ kN/m}$$

Summation of moments about point C gives the location of the resultant active force against the retaining wall.

$$y = \frac{R_1 y_1 + R_2 y_2 + R_3 y_3 + R_4 y_4 + R_5 y_5}{R_1 + R_2 + R_3 + R_4 + R_5}$$

$$= \frac{\begin{matrix} \left(27.1 \frac{\text{kN}}{\text{m}} \right)(8.25 \text{ m}) + \left(37.0 \frac{\text{kN}}{\text{m}} \right)(7.67 \text{ m}) \\ + \left(187.9 \frac{\text{kN}}{\text{m}} \right)(3.25 \text{ m}) \\ + \left(68.6 \frac{\text{kN}}{\text{m}} \right)(2.17 \text{ m}) \\ + \left(207.4 \frac{\text{kN}}{\text{m}} \right)(2.17 \text{ m}) \end{matrix}}{\begin{matrix} 27.1 \frac{\text{kN}}{\text{m}} + 37.0 \frac{\text{kN}}{\text{m}} + 187.9 \frac{\text{kN}}{\text{m}} \\ + 68.6 \frac{\text{kN}}{\text{m}} + 207.4 \frac{\text{kN}}{\text{m}} \end{matrix}}$$

$$= 3.25 \text{ m} \quad (3.3 \text{ m})$$

The answer is (B).

89. For a loose, natural sand deposit, the coefficient of earth pressure at rest can be estimated as

$$k_0 \approx 1 - \sin \phi = 1 - \sin 29° = 0.52$$

The total at-rest lateral earth pressure is

$$p_o = k_o p_v$$

This can be rewritten as

$$\sigma_h = \sigma_h' + u = k_o \sigma_v' + u = k_o \gamma_b z + u$$
$$= k_o (\gamma_{sat} - \gamma_w) z + \gamma_w z$$
$$= (0.52) \left(19.3 \, \frac{kN}{m^3} - 9.81 \, \frac{kN}{m^3} \right) (10 \text{ m})$$
$$+ \left(9.81 \, \frac{kN}{m^3} \right) (10 \text{ m})$$
$$= 147.4 \text{ kN/m}^2 \quad (150 \text{ kPa})$$

The answer is (B).

90. Darcy's law gives the effective (apparent) velocity through the clay liner.

$$v_e = Ki = K \frac{h_L}{L}$$
$$= \left(2.5 \times 10^{-7} \, \frac{mm}{s} \right) \left(\frac{2.25 \text{ m}}{1.00 \text{ m}} \right)$$
$$= 5.63 \times 10^{-7} \text{ mm/s}$$

The problem asks for pore velocity, v_{pore}. Porosity, n, relates the pore velocity to the effective velocity of the flow.

$$e = \frac{n}{1 - n}$$
$$n = \frac{e}{1 + e} = \frac{0.88}{1 + 0.88}$$
$$= 0.47$$

$$v_{pore} = \frac{v_e}{n}$$
$$= \frac{5.63 \times 10^{-7} \, \frac{mm}{s}}{0.47}$$
$$= 1.2 \times 10^{-6} \text{ mm/s}$$

The answer is (C).

91. This problem is solved using the standard weight-volume relationships for soils.

The moisture content is

$$w = \frac{W_w}{W_s} = \frac{m_w}{m_s}$$
$$= \frac{1733 \text{ g} - 1287 \text{ g}}{1287 \text{ g}} \times 100\%$$
$$= 34.7\%$$

Since the soil sample is saturated, the degree of saturation is 100%. The void ratio is

$$Se = wSG$$
$$e = \frac{wSG}{S} = \frac{(34.7\%)(2.7)}{100\%}$$
$$= 0.937$$

The total unit weight is

$$\gamma = \left(\frac{SG + Se}{1 + e} \right) \gamma_w$$
$$= \left(\frac{2.7 + 0.937}{1 + 0.937} \right) \left(9.81 \, \frac{kN}{m^3} \right)$$
$$= 18.4 \text{ kN/m}^3$$

The answer is (D).

92. Since the backfill is horizontal and the retaining wall is smooth, the coefficient of active earth pressure is

$$k_a = \tan^2 \left(45° - \frac{\phi}{2} \right)$$
$$= \tan^2 \left(45° - \frac{30°}{2} \right)$$
$$= 0.33$$

The active force per unit length can be calculated using Rankine theory.

$$P_a = \tfrac{1}{2} \gamma H^2 k_a$$
$$= \left(\frac{1}{2} \right) \left(20 \, \frac{kN}{m^3} \right) (2.2 \text{ m})^2 (0.33)$$
$$= 16.0 \text{ kN/m}$$

The overturning moment per unit length of wall is given by the following equation. M is the overturning moment per unit length of wall, and M_B is the total overturning moment.

$$M = \frac{M_B}{L} = P_a \frac{H}{3}$$
$$= \left(16.0 \, \frac{kN}{m} \right) \left(\frac{2.2 \text{ m}}{3} \right)$$
$$= 11.7 \text{ kN·m/m}$$

The weight of the wall (per unit length of wall) is the only vertical force.

$$W = \frac{P}{L} = HB\gamma_{concrete}$$
$$= (2.8 \text{ m})(1.0 \text{ m}) \left(25 \, \frac{kN}{m^3} \right)$$
$$= 70 \text{ kN/m}$$

The vertical pressure at point A is the maximum pressure beneath the wall.

$$p_A = p_{max} = \left(\frac{P}{BL}\right)\left(1 + \frac{6\epsilon}{B}\right)$$

$$= \left(\frac{W}{B}\right)\left(1 + \frac{6\epsilon}{B}\right)$$

The eccentricity is

$$\epsilon = \frac{M_B}{P} = \frac{M}{W}$$

$$= \frac{11.7 \; \frac{\text{kN·m}}{\text{m}}}{70 \; \frac{\text{kN}}{\text{m}}}$$

$$= 0.167 \text{ m}$$

$$p_A = \frac{W}{B}\left(1 + \frac{6\epsilon}{B}\right)$$

$$= \left(\frac{70 \; \frac{\text{kN}}{\text{m}}}{1.0 \text{ m}}\right)\left(1 + \frac{(6)(0.167 \text{ m})}{1.0 \text{ m}}\right)$$

$$= 140 \text{ kN/m}^2 \quad (140 \text{ kPa})$$

The answer is (D).

93. For uniform sands, the permeability can be estimated using the following equation.

$$K_{\text{mm/s}} \approx CD_{10,\text{mm}}^2$$

The coefficient C varies between 10 and 15. The effective grain size for this soil is approximately

$$D_{10,\text{mm}} = 0.15 \text{ mm}$$

Assuming C equals 12,

$$K_{\text{mm/s}} \approx (12)(0.15)^2$$

$$\approx 0.27 \text{ mm/s} \quad (3.0 \times 10^{-3} \text{ cm/s})$$

The answer is (B).

94. The different vertical coefficients of permeability in these stratified anisotropic soils can be combined into one effective vertical coefficient of permeability, K.

$$K = \frac{\sum H_j}{\sum \frac{H_j}{K_j}}$$

$$= \frac{2.5 \text{ m} + 1.0 \text{ m} + 3.0 \text{ m}}{\frac{2.5 \text{ m}}{3.0 \times 10^{-5} \; \frac{\text{mm}}{\text{s}}} + \frac{1.0 \text{ m}}{2.0 \times 10^{-6} \; \frac{\text{mm}}{\text{s}}}}$$

$$+ \frac{3.0 \text{ m}}{3.8 \times 10^{-6} \; \frac{\text{mm}}{\text{s}}}$$

$$= 4.7 \times 10^{-6} \text{ mm/s}$$

From Darcy's Law,

$$V = Qt = KiAt$$

$$= K\frac{h_L}{L}At$$

$$= \left(4.7 \times 10^{-6} \; \frac{\text{mm}}{\text{s}}\right)\left(\frac{1 \text{ m}}{1000 \text{ mm}}\right)$$

$$\times \left(\frac{1.5 \text{ m}}{2.5 \text{ m} + 1.0 \text{ m} + 3.0 \text{ m}}\right)$$

$$\times (5000 \text{ m}^2)(6 \text{ mo})\left(\frac{30 \text{ d}}{1 \text{ mo}}\right)$$

$$\times \left(\frac{24 \text{ h}}{1 \text{ d}}\right)\left(\frac{3600 \text{ s}}{1 \text{ h}}\right)$$

$$= 84.3 \text{ m}^3 \quad (85 \text{ m}^3)$$

The answer is (A).

95. From the elevation information given, the total head is

$$h = 365 \text{ m} - 360 \text{ m} = 5 \text{ m}$$

From the flow net, the total number of head drops, N_D, is 14, and the number of head drops to point A is 12. The depth at point A is

$$z_A = 360 \text{ m} - 350 \text{ m}$$

$$= 10 \text{ m}$$

The head lost at point A is

$$\Delta h_A = h\frac{\text{number of head drops to A}}{N_D}$$

$$= (5 \text{ m})\left(\frac{12}{14}\right)$$

$$= 4.29 \text{ m}$$

The pore pressure at point A is

$$p_A = \gamma_w(h + z_A - \Delta h_A)$$

$$= \left(9.81 \; \frac{\text{kN}}{\text{m}^3}\right)(5 \text{ m} + 10 \text{ m} - 4.29 \text{ m})$$

$$= 105 \text{ kN/m}^2 \quad (105 \text{ kPa})$$

The answer is (B).

96. The effective area is the greatest possible portion of the footing such that the resultant force passes through its centroid. Given a rectangular footing of width B and length L, and a load with eccentricity ϵ of 0.15 m in the

B-direction and 0.5 m in the L-direction, the equivalent width B' and length L' of the effective area are

$$B' = B - 2\epsilon_B$$
$$= 1.5 \text{ m} - (2)(0.15 \text{ m})$$
$$= 1.2 \text{ m}$$
$$L' = L - 2\epsilon_L$$
$$= 3.0 \text{ m} - (2)(0.5 \text{ m})$$
$$= 2.0 \text{ m}$$

The effective area is

$$A_e = B'L'$$
$$= (1.2 \text{ m})(2.0 \text{ m})$$
$$= 2.4 \text{ m}^2$$

The answer is (C).

97. The ultimate bearing capacity is given by the following equation.

$$q_{\text{ult}} = c\lambda_{cs}\lambda_{cd}N_c + q\lambda_{qs}\lambda_{qd}N_q + \frac{1}{2}\lambda_{\gamma s}\lambda_{\gamma d}\gamma B N_\gamma$$

From a table of Terzaghi bearing capacity factors, when ϕ equals $0°$, then N_c is 5.7, N_q is 1.0, and N_γ is 0.

The shape and depth factors are

$$\lambda_{qs} = \lambda_{\gamma s} = 1$$
$$\lambda_{qd} = \lambda_{\gamma d} = 1$$

The value of B/L is so small it can be taken as zero, so

$$\lambda_{cs} = 1 + 0.2\frac{B}{L}\tan^2\left(45° + \frac{\phi}{2}\right)$$
$$= 1$$
$$\lambda_{cd} = 1 + 0.2\frac{D_f}{B}\tan\left(45° + \frac{\phi}{2}\right)$$
$$= 1 + (0.2)\left(\frac{1.0 \text{ m}}{2.0 \text{ m}}\right)\tan\left(45° + \frac{0°}{2}\right)$$
$$= 1.1$$

The surcharge is

$$q = (\gamma_{\text{sat}} - \gamma_w)D_f$$
$$= \left(18.5 \frac{\text{kN}}{\text{m}^3} - 9.8 \frac{\text{kN}}{\text{m}^3}\right)(1.0 \text{ m})$$
$$= 8.7 \text{ kN/m}^2$$

Therefore,

$$q_{\text{ult}} = c\lambda_{cd}N_c + q$$
$$= \left(110 \frac{\text{kN}}{\text{m}^2}\right)(1.1)(5.7) + 8.7 \frac{\text{kN}}{\text{m}^2}$$
$$= 698.4 \text{ kN/m}^2$$
$$P_{\text{ult}} = q_{\text{ult}}B$$
$$= \left(698.4 \frac{\text{kN}}{\text{m}^2}\right)(2.0 \text{ m})$$
$$= 1397 \text{ kN/m} \quad (1400 \text{ kN/m})$$

The answer is (D).

98. The rock quality designation is defined as the total length of all intact pieces 10 cm or longer, divided by the total length of the core, expressed as a percentage.

$$\text{RQD} = \left(\frac{89 \text{ cm}}{123 \text{ cm}}\right) \times 100\%$$
$$= 72\%$$

The answer is (C).

99. The ultimate pullout capacity of the pile is composed of the pile weight and the skin friction. The weight of the pile is

$$W = LA\gamma_c = L\frac{\pi d^2}{4}\gamma_c$$
$$= (6.0 \text{ m})\left(\frac{\pi(0.30 \text{ m})^2}{4}\right)\left(25 \frac{\text{kN}}{\text{m}^3}\right)$$
$$= 10.6 \text{ kN}$$

The skin friction area is πdL. The skin friction is

$$Q_f = k_h\sigma_v'\tan\delta\pi dL$$
$$= k_h\gamma\frac{L}{2}\tan\delta\pi dL$$
$$= (1.1)\left(20 \frac{\text{kN}}{\text{m}^3}\right)\left(\frac{6.0 \text{ m}}{2}\right)(\tan 25°)$$
$$\times \pi(0.30 \text{ m})(6.0 \text{ m})$$
$$= 174.0 \text{ kN}$$

σ_v' is the average effective vertical pressure of the soil along the pipe and is equal to $\gamma L/2$.

The pullout capacity is

$$P = W + Q_f = 10.6 \text{ kN} + 174.0 \text{ kN}$$
$$= 184.6 \text{ kN} \quad (180 \text{ kN})$$

The answer is (C).

100. The total primary consolidation settlement is

$$S_p = \frac{C_c H_c}{1 + e_o} \log\left(\frac{p'_o + \Delta p'_v}{p'_o}\right)$$
$$= \frac{(0.26)(8 \text{ m})}{1 + 1.02} \log\left(\frac{240 \text{ kPa} + 130 \text{ kPa}}{240 \text{ kPa}}\right)$$
$$= 0.194 \text{ m} \quad (200 \text{ mm})$$

The answer is (D).

101. The time factor for the clay layer is

$$T_v = \frac{C_v t}{H_d^2}$$
$$= \frac{\left(7.6 \times 10^{-8} \, \frac{\text{m}^2}{\text{s}}\right)(3 \text{ yr})\left(\frac{365 \text{ d}}{1 \text{ yr}}\right)}{\left(\frac{6.0 \text{ m}}{2}\right)^2}$$
$$\times \left(\frac{24 \text{ h}}{1 \text{ d}}\right)\left(\frac{3600 \text{ s}}{1 \text{ h}}\right)}{\left(\frac{6.0 \text{ m}}{2}\right)^2}$$
$$= 0.80$$

Interpolating from a table of time factors, the degree of consolidation, U_z, for a time factor of 0.80 is about 88%, or most nearly 90%.

The answer is (D).

102. The time factor, T_v, for a specified average degree of consolidation, U_z, is found from a table of approximate time factors.

For an average degree of consolidation of 0.90, T_v is approximately equal to 0.85.

The time required to achieve this degree of consolidation is

$$t = \frac{T_v H_d^2}{C_v}$$
$$= \frac{(0.85)\left(\frac{20 \text{ m}}{2}\right)^2}{\left(4.3 \times 10^{-7} \, \frac{\text{m}^2}{\text{s}}\right)\left(\frac{3600 \text{ s}}{1 \text{ h}}\right)\left(\frac{24 \text{ h}}{1 \text{ d}}\right)\left(\frac{365 \text{ d}}{1 \text{ yr}}\right)}$$
$$= 6.3 \text{ yr}$$

The answer is (C).

103. For one-dimensional loading, the excess pore water pressure at beginning of loading, u_i, is 100 kPa. As time passes, the excess pore water pressure decreases.

The answer is (D).

104. Two formulas for calculating dry unit weight are

$$\gamma_d = \frac{\gamma}{1 + w}$$
$$\gamma_d = \frac{SG\gamma_w}{1 + e}$$

Therefore,

$$\frac{\gamma}{1 + w} = \frac{SG\gamma_w}{1 + e}$$
$$e = \frac{SG(1 + w)\gamma_w}{\gamma} - 1$$
$$= \frac{(2.72)(1 + 0.08)\left(9.81 \, \frac{\text{kN}}{\text{m}^3}\right)}{17.6 \, \frac{\text{kN}}{\text{m}^3}} - 1$$
$$= 0.64$$

The degree of saturation is

$$S = \frac{SGw}{e} = \frac{(2.72)(0.08)}{0.64} \times 100\%$$
$$= 34\%$$

The answer is (D).

105. From the grain size distribution, D_{10} is 0.02 mm, D_{30} is 0.6 mm, and D_{60} is 8.5 mm.

The coefficient of curvature is

$$C_c = \frac{D_{30}^2}{D_{10}D_{60}}$$
$$= \frac{(0.6 \text{ mm})^2}{(0.02 \text{ mm})(8.5 \text{ mm})}$$
$$= 2.1 \quad (2.0)$$

The answer is (A).

106. In order to determine the appropriate active soil pressure envelope to use, it is necessary to determine if the clay is soft, medium, or stiff by calculating the stability number, N_o.

$$N_o = \frac{\gamma H}{c} = \frac{\left(18.3 \, \frac{\text{kN}}{\text{m}^3}\right)(8 \text{ m})}{23 \, \frac{\text{kN}}{\text{m}^2}}$$
$$= 6.4$$

N_o is greater than 6, so the braced cut is in soft clay. The pressure distribution for soft clay is shown.

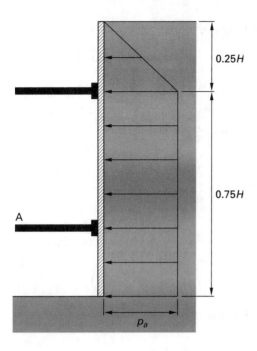

In this pressure distribution (which is strictly applicable for $H \geq 6$ m and a water table below the bottom of the cut), active pressure is

$$p_a = \gamma H - 4c$$
$$= \left(18.3 \; \frac{\text{kN}}{\text{m}^3}\right)(8 \text{ m}) - (4)\left(23 \; \frac{\text{kN}}{\text{m}^2}\right)$$
$$= 54.4 \text{ kN/m}^2 \quad (54.4 \text{ kPa})$$

The pressure distribution is

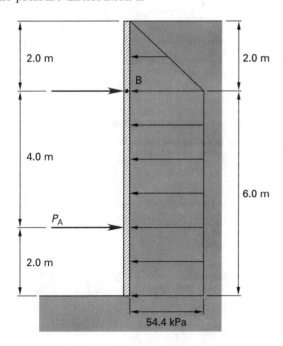

By resolving the pressure distribution into resultant forces and taking moments about point B, the force on strut A is

$$\sum M_{\text{B}} = P_{\text{A}}(4.0 \text{ m}) - \left(54.4 \; \frac{\text{kN}}{\text{m}^2}\right)(6.0 \text{ m})$$
$$\times \left(\frac{6.0 \text{ m}}{2}\right) - \left(\frac{1}{2}\right)\left(54.4 \; \frac{\text{kN}}{\text{m}^2}\right)$$
$$\times (2.0 \text{ m})\left(\frac{2.0 \text{ m}}{3}\right)(4.0 \text{ m})$$
$$= 0$$

$$P_{\text{A}} = \frac{\left(54.4 \; \frac{\text{kN}}{\text{m}^2}\right)(6.0 \text{ m})\left(\frac{6.0 \text{ m}}{2}\right)}{4.0 \text{ m}}$$
$$\frac{- \left(\frac{1}{2}\right)\left(54.4 \; \frac{\text{kN}}{\text{m}^2}\right)(2.0 \text{ m})\left(\frac{2.0 \text{ m}}{3}\right)}{4.0 \text{ m}}$$
$$= 943 \text{ kN} \quad (940 \text{ kN})$$

The answer is (C).

107. The unit skin friction along the pile is found by the equation

$$f_s = k\sigma_v' \tan \delta$$

The critical depth is $15B$. For a depth, z, from 0 to $15B$,

$$\sigma_v' = \gamma z = 18.8z$$

For a depth greater or equal to $15B$,

$$\sigma_v' = \gamma z = \gamma 15B$$
$$= \left(18.8 \; \frac{\text{kN}}{\text{m}^3}\right)(15)(0.254 \text{ m})$$
$$= 71.6 \text{ kN/m}^2 \quad (71.6 \text{ kPa})$$

The resulting vertical effective stress distribution is shown.

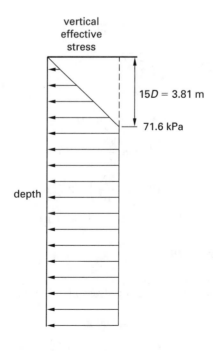

For the four sides of the square pile $(4BL')$ and the average pressure, the skin-friction capacity along the pile from depth 0 to $15D$ is

$$Q_f = 4BL'f_{\text{ave}} = 4BL'\tfrac{1}{2}k\sigma'_v \tan \delta$$

$$= (4)\,(0.254 \text{ m})\,(15)\,(0.254 \text{ m})\left(\frac{1}{2}\right)$$

$$\times (1.6)\left(71.6 \ \frac{\text{kN}}{\text{m}^2}\right)(\tan 0.6 \times 35°)$$

$$= 85.1 \text{ kN}$$

The skin-friction capacity along the pile from depth $15D$ to 10 m is

$$Q_f = 4B\,(L - L')\,f_{z=15D}$$

$$= (4)\,(0.254 \text{ m})\,(10 \text{ m} - 3.81 \text{ m})$$

$$\times (1.6)\left(71.6 \ \frac{\text{kN}}{\text{m}^2}\right)\tan\,(0.6 \times 35°)$$

$$= 276.6 \text{ kN}$$

The total frictional resistance along the pile is

$$Q_{s,\text{total}} = \sum Q_s = 85.1 \text{ kN} + 276.6 \text{ kN}$$

$$= 362 \text{ kN} \quad (360 \text{ kN})$$

The answer is (C).

108. Rankine theory for a passive earth condition should be used to solve this problem. The passive earth coefficient is

$$k_p = \tan^2\left(45° + \frac{\phi}{2}\right)$$

$$= \tan^2\left(45° + \frac{28°}{2}\right)$$

$$= 2.77$$

The total force per meter of wall is

$$P = \tfrac{1}{2}\gamma H^2 k_p + 2c\sqrt{k_p}H$$

$$= \left(\frac{1}{2}\right)\left(18.8 \ \frac{\text{kN}}{\text{m}^3}\right)(3.0 \text{ m})^2\,(2.77)$$

$$+ (2)\left(16 \ \frac{\text{kN}}{\text{m}^2}\right)\sqrt{2.77}\,(3.0 \text{ m})$$

$$= 394.1 \text{ kN/m} \quad (395 \text{ kN/m})$$

The answer is (B).

109. The effective minor stress at failure is

$$\overline{\sigma}_3 = 200 \text{ kPa}$$

The effective major stress at failure is

$$\overline{\sigma}_1 = 200 \text{ kPa} + 468 \text{ kPa}$$

$$= 668 \text{ kPa}$$

The angle of internal friction can be calculated using the following equation.

$$\frac{\overline{\sigma}_1}{\overline{\sigma}_3} = \frac{1 + \sin \phi}{1 - \sin \phi}$$

$$\sin \phi = \frac{\overline{\sigma}_1 - \overline{\sigma}_3}{\overline{\sigma}_1 + \overline{\sigma}_3}$$

$$= \frac{668 \text{ kPa} - 200 \text{ kPa}}{668 \text{ kPa} + 200 \text{ kPa}}$$

$$= 0.539$$

$$\phi = 32.6° \quad (33°)$$

The answer is (C).

110. The buoyant unit weight is

$$\gamma_b = \gamma_{\text{sat}} - \gamma_w$$

$$= \left(\frac{\text{SG} + e}{1 + e}\right)\gamma_w - \gamma_w$$

$$= \left(\frac{2.66 + 0.62}{1 + 0.62}\right)\left(9.81 \ \frac{\text{kN}}{\text{m}^3}\right) - \left(9.81 \ \frac{\text{kN}}{\text{m}^3}\right)$$

$$= 10.1 \text{ kN/m}^3 \quad (10 \text{ kN/m}^3)$$

The answer is (A).

111. The ultimate bearing capacity is given by the equation

$$q_{\text{ult}} = cN_c + (p_q + \gamma D_f)N_q + \tfrac{1}{2}\gamma BN_\gamma$$

The footing is placed near the ground surface, and p_q equals 0 kN/m^2, so $p_q + \gamma D_f$ is zero.

$$q_{\text{ult}} = cN_c + \tfrac{1}{2}\gamma BN_\gamma$$

A table of Terzaghi bearing capacity factors shows that when ϕ equals 25°, N_c equals 25.1, and N_γ equals 9.7.

$$q_{\text{ult}} = cN_c + \tfrac{1}{2}\gamma BN_\gamma$$

$$= \left(14 \ \frac{\text{kN}}{\text{m}^2}\right)(25.1) + \left(\frac{1}{2}\right)\left(18.6 \ \frac{\text{kN}}{\text{m}^3}\right)$$

$$\times (1.5 \text{ m})\,(9.7)$$

$$= 486.7 \text{ kN/m}^2$$

$$q_{\text{actual}} = \frac{P_{\text{actual}}}{B} = \frac{596 \ \dfrac{\text{kN}}{\text{m}}}{1.5 \text{ m}}$$

$$= 397.3 \text{ kN/m}^2$$

$$FS = \frac{q_{ult}}{q_{actual}} = \frac{486.7 \ \frac{kN}{m^2}}{397.3 \ \frac{kN}{m^2}}$$

$$= 1.22 \quad (1.2)$$

The answer is (B).

112. Coulomb's equation relates soil strength to the normal stress on the failure plane.

$$S = \tau = c + \sigma \tan \phi$$

The shear strength of sand, c, is 0, so

$$\tau = \sigma \tan \phi$$

$$\tan \phi = \frac{\tau}{\sigma} = \frac{63.4 \ \frac{kN}{m^2}}{100 \ \frac{kN}{m^2}}$$

$$= 0.634$$

Use the same relationship to calculate the shear strength of that soil for a normal stress of 75 kN/m².

$$\tau = \sigma \tan \phi = \left(75 \ \frac{kN}{m^2} \right) (0.634)$$

$$= 47.6 \ kN/m^2$$

The shear force required to cause failure is

$$S = \tau A = \left(47.6 \ \frac{kN}{m^2} \right) (0.06 \ m)^2$$

$$= 0.17 \ kN$$

The answer is (A).

113. During an undrained test, the volume of the sample does not change.

$$V = HA = H_0 \frac{\pi d_0^2}{4}$$

$$= H_f \frac{\pi d_f^2}{4}$$

$$d_f = \sqrt{\frac{H_0}{H_f}} d_0$$

$$= \sqrt{\frac{9.1 \ cm}{8.67 \ cm}} (4.0 \ cm)$$

$$= 4.1 \ cm \quad (0.041 \ m)$$

The maximum principal stress is

$$\sigma_1 = \frac{P}{A_f} = \frac{P}{\frac{\pi d_f^2}{4}}$$

$$= \frac{0.43 \ kN}{\frac{\pi (0.041 \ m)^2}{4}}$$

$$= 326 \ kN/m^2$$

The undrained shear strength is

$$S_u = c = \frac{\sigma_1}{2}$$

$$= \frac{326 \ \frac{kN}{m^2}}{2}$$

$$= 163 \ kN/m^2 \quad (160 \ kPa)$$

The answer is (B).

114. Since this test is performed under drained conditions, there is no pore water pressure.

$$\sigma_1' = \sigma_3' + \Delta\sigma_{D_f}$$

$$= 280 \ \frac{kN}{m^2} + 410 \ \frac{kN}{m^2}$$

$$= 690 \ kN/m^2$$

The clay has no drained cohesion, so the effective principal stresses at failure can be related by this equation, solving for ϕ.

$$\frac{\sigma_1'}{\sigma_3'} = \frac{1 + \sin \phi}{1 - \sin \phi}$$

The following equation is also commonly used.

$$\sigma_1' = \sigma_3' \tan^2 \left(45° + \frac{\phi}{2} \right)$$

Solving for ϕ,

$$\phi = 2 \arctan \sqrt{\frac{\sigma_1'}{\sigma_3'}} - 45°$$

$$= 2 \arctan \sqrt{\frac{690 \ \frac{kN}{m^2}}{280 \ \frac{kN}{m^2}}} - 45°$$

$$= 25°$$

The angle between the failure plane and the major principal plane is

$$\alpha = 45° + \frac{\phi}{2}$$
$$= 45° + \frac{25°}{2}$$
$$= 57.5°$$

The shear stress on the failure plane is

$$\tau_f = \frac{\sigma_1' - \sigma_3'}{2} \sin 2\theta$$
$$= \left(\frac{690 \ \frac{kN}{m^2} - 280 \ \frac{kN}{m^2}}{2} \right) \sin \left((2)(57.5°) \right)$$
$$= 185.8 \ kN/m^2 \quad (190 \ kPa)$$

The answer is (A).

115. The void ratio of the sand can be calculated using the following formula.

$$e = \frac{SG\gamma_w}{\gamma_d} - 1$$
$$= \frac{(2.65) \left(9.81 \ \frac{kN}{m^3} \right)}{16.5 \ \frac{kN}{m^3}} - 1$$
$$= 0.576$$

Relative density can be calculated with the following formula.

$$D_r = \frac{e_{max} - e}{e_{max} - e_{min}}$$
$$= \frac{0.78 - 0.576}{0.78 - 0.41}$$
$$= 0.55$$

The answer is (B).

116. The group index is computed as

$$I_g = (F_{200} - 35)\left(0.2 + 0.005(LL - 40)\right)$$
$$\quad + (0.01)(F_{200} - 15)(PI - 10)$$
$$= (45 - 35)\left(0.2 + (0.005)(40 - 40)\right)$$
$$\quad + (0.01)(45 - 15)(13 - 10)$$
$$= 2.9 \quad (3)$$

The percentage passing a no. 200 sieve is greater than 36%, the liquid limit is 40, and the plasticity index is

greater than 11. The AASHTO classification and group index of this soil is A-6 (3).

The answer is (C).

117. Use a USCS table to solve this problem.

As less than 50% passes a no. 200 sieve, the soil is coarse grained.

More than half the coarse fraction is finer than no. 4, so the soil is a sand (S).

More than 12% passes a no. 200 sieve, so the soil will be classified as either SM or SC.

The plasticity index is

$$PI = LL - PL$$
$$= 55 - 20$$
$$= 35$$

With a liquid limit of 55 and a plasticity index of 35, the fine-grained fraction of the soil classifies as highly plastic clay (CH).

The soil is therefore classified as SC.

The answer is (A).

118. The active force can be calculated using the Rankine theory.

$$R_a = \frac{1}{2}\gamma H^2 k_a$$
$$= \left(\frac{1}{2} \right) \left(20 \ \frac{kN}{m^3} \right) (5.5 \ m)^2 (0.33)$$
$$= 99.8 \ kN/m$$

The backfill is horizontal, and the wall face is vertical.

$$k_a = \tan^2 \left(45° - \frac{\phi}{2} \right)$$
$$= \tan^2 \left(45° - \frac{30°}{2} \right)$$
$$= 0.33$$

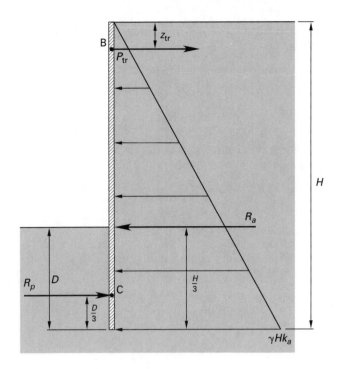

All forces are in equilibrium, so the sum of the moments about point B is zero. Taking moments about B allows the tie rod force to be ignored. The moment balance is

$$\sum M_B = R_a \left(\tfrac{2}{3}H - z_{tr}\right) - R_p \left(H - z_{tr} - \tfrac{1}{3}D\right)$$
$$= 0$$

$$R_p = R_a \left(\frac{\tfrac{2}{3}H - z_{tr}}{H - z_{tr} - \tfrac{1}{3}D}\right)$$

$$= \left(99.8 \ \frac{kN}{m}\right)\left(\frac{\left(\tfrac{2}{3}\right)(5.5 \text{ m}) - (0.5 \text{ m})}{5.5 \text{ m} - 0.5 \text{ m} - \left(\tfrac{1}{3}\right)(2.1 \text{ m})}\right)$$

$$= 73.5 \text{ kN/m} \quad (70 \text{ kN/m})$$

The answer is (A).

119. To have a fully compensated foundation, the total weight of the excavated soil must be equal to the total load on the foundation.

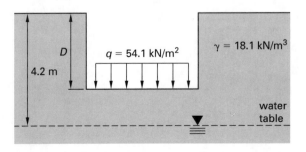

The pressure due to the total dead and live load is

$$q = \frac{P}{BL} = \frac{33\,540 \text{ kN}}{(20 \text{ m})(31 \text{ m})}$$
$$= 54.1 \text{ kN/m}^2$$

The embedment depth required to have a fully compensated foundation is

$$\gamma D = q$$

$$D = \frac{q}{\gamma} = \frac{54.1 \ \dfrac{kN}{m^2}}{18.1 \ \dfrac{kN}{m^3}}$$

$$= 3.0 \text{ m}$$

The answer is (C).

120. The factor of safety against sliding can be expressed as

$$FS_{sl} = \frac{\sum R_r}{\sum R_o}$$

$\sum R_r$ is the sum of forces per unit length of wall resisting sliding, and $\sum R_o$ is the sum of forces per unit length of wall causing sliding.

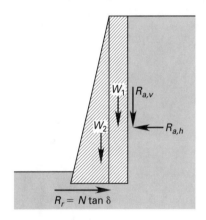

The active force is

$$R_a = \tfrac{1}{2}\gamma H^2 k_a$$
$$= \left(\tfrac{1}{2}\right)\left(20.2 \ \frac{kN}{m^3}\right)(4.3 \text{ m})^2 (0.31)$$
$$= 57.9 \text{ kN/m}$$

The horizontal and vertical components of the active force are

$$R_{a,h} = R_a \cos \delta$$
$$= \left(57.9 \ \frac{kN}{m}\right)\cos 20°$$
$$= 54.4 \text{ kN/m}$$

$$R_{a,v} = R_a \sin \delta$$

$$= \left(57.9 \ \frac{\text{kN}}{\text{m}}\right) \sin 20°$$

$$= 19.8 \ \text{kN/m}$$

The weight of the wall can be subdivided into W_1 and W_2.

$$W_1 = H b_1 \gamma_{\text{concrete}}$$

$$= (4.3 \ \text{m})(0.4 \ \text{m}) \left(25 \ \frac{\text{kN}}{\text{m}^3}\right)$$

$$= 43.0 \ \text{kN/m}$$

$$W_2 = \tfrac{1}{2} H b_2 \gamma_{\text{concrete}}$$

$$= \left(\frac{1}{2}\right)(4.3 \ \text{m})(2.1 \ \text{m}) \left(25 \ \frac{\text{kN}}{\text{m}^3}\right)$$

$$= 112.9 \ \text{kN/m}$$

$$W = W_1 + W_2$$

$$= 43.0 \ \frac{\text{kN}}{\text{m}} + 112.9 \ \frac{\text{kN}}{\text{m}}$$

$$= 155.9 \ \text{kN/m}$$

The total normal force acting on the base of the wall is

$$N = W + R_{a,v}$$

$$= 155.9 \ \frac{\text{kN}}{\text{m}} + 19.8 \ \frac{\text{kN}}{\text{m}}$$

$$= 175.7 \ \text{kN/m}$$

The only force that resists sliding per unit length of wall is

$$R_r = N \tan \delta$$

$$= \left(175.9 \ \frac{\text{kN}}{\text{m}}\right) \tan 20°$$

$$= 63.9 \ \text{kN/m}$$

The horizontal component of the active force is the only force that tends to slide the wall.

$$R_o = P_{a,h}$$

$$= 54.4 \ \text{kN/m}$$

The factor of safety against sliding is

$$\text{FS}_{\text{sl}} = \frac{R_r}{R_o} = \frac{63.9 \ \frac{\text{kN}}{\text{m}}}{54.4 \ \frac{\text{kN}}{\text{m}}}$$

$$= 1.17 \quad (1.2)$$

The answer is (B).

Solutions
Structural

121. The resultant of the three wheel loads for an HS20-44 loading is a 36 kips force located 4.67 ft from the 16 kips center force.

$$\overline{x} = \frac{\sum P_i x_i}{R}$$
$$= \frac{(16 \text{ kips})(14 \text{ ft}) + (4 \text{ kips})(28 \text{ ft})}{36 \text{ kips}}$$
$$= 9.33 \text{ ft}$$

Maximum wheel-load bending moment occurs when the midspan lies halfway between the resultant and the central 16 kips force. Thus, the position for maximum wheel-load bending moment is

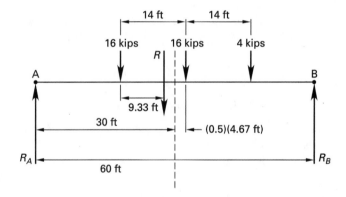

Maximum wheel-load bending moment occurs under the 16 kips load to the right of midspan.

$$R_\text{A} = \frac{\sum rF}{L}$$
$$= \frac{(36 \text{ kips})\big(30 \text{ ft} + (0.5)(4.67 \text{ ft})\big)}{60 \text{ ft}}$$
$$= 19.4 \text{ kips}$$

$$M_\text{max} = \sum rF = (19.4 \text{ kips})\big(30 \text{ ft} + (0.5)(4.67 \text{ ft})\big)$$
$$- (16 \text{ kips})(14 \text{ ft})$$
$$= 403 \text{ ft-kips}$$

The AASHTO specification requires an increase in the wheel-load bending moment to account for impact, and a distribution factor to the individual girder that is based on the girder spacing.

$$I = \frac{50 \text{ ft}}{L + 125} = \frac{50 \text{ ft}}{60 \text{ ft} + 125 \text{ ft}}$$
$$= 0.27$$

$$\text{DF} = \frac{S}{5.5 \text{ ft}} = \frac{7.0 \text{ ft}}{5.5 \text{ ft}}$$
$$= 1.27$$

$$M_L = (\text{DF})M_\text{max}$$
$$= (1.27)(403 \text{ ft-kips})$$
$$= 512 \text{ ft-kips}$$

$$M_u = 1.3\Big(M_D + \frac{5}{3}(1 + I)M_L\Big)$$
$$= (1.3)(500 \text{ ft-kips}$$
$$+ \Big(\frac{5}{3}\Big)(1 + 0.27)(512 \text{ ft-kips}))$$
$$= 2058 \text{ ft-kips} \quad (2100 \text{ ft-kips})$$

The answer is (D).

122. The resultant lateral force is
$$V = wL = \Big(0.4 \frac{\text{kips}}{\text{ft}}\Big)(160 \text{ ft})$$
$$= 64 \text{ kips}$$

This resultant force acts 80 ft from the west wall. The center of rigidity of the wall group is

$$\overline{x} = \frac{\sum R_i x_i}{\sum R_i}$$
$$= \frac{(4R)(0 \text{ ft}) + (3R)(120 \text{ ft}) + (3R)(160 \text{ ft})}{4R + 3R + 3R}$$
$$= 84 \text{ ft} \quad [\text{from the west side of wall A}]$$

From symmetry
$$\overline{y} = 30 \text{ ft} \quad [\text{from the south wall}]$$

The wall system is subjected to a torsional moment of
$$M_t = V\Big(\overline{x} - \frac{L}{2}\Big)$$
$$= (64 \text{ kips})\Big(84 \text{ ft} - \frac{160 \text{ ft}}{2}\Big)$$
$$= 256 \text{ ft-kips clockwise}$$

The polar moment of inertia for the walls resisting the torsional moment is

$$J = \sum(R_{yi}x_i^2 + R_{xi}y_i^2)$$
$$= (4R)(-84 \text{ ft})^2 + (3R)(120 \text{ ft} - 84 \text{ ft})^2$$
$$+ (3R)(160 \text{ ft} - 84 \text{ ft})^2$$
$$+ (R)(30 \text{ ft})^2 + (R)(30 \text{ ft})^2$$
$$= 51{,}240R \text{ ft}^2$$

The maximum lateral force resisted by wall A is the combined direct force plus the force caused by the torsional moment, both acting in the same sense.

$$V_A = \frac{4R}{\sum R_{yi}}V + \frac{M_t R_i x_i}{J}$$
$$= \frac{4R}{10R}(64 \text{ kips}) + \frac{(256 \text{ ft-kips})(4R)(84 \text{ ft})}{51{,}240R \text{ ft}^2}$$
$$= 27.3 \text{ kips} \quad (27 \text{ kips})$$

The answer is (C).

123. The plywood diaphragm is considered flexible, and the lateral forces transfer to the shearwall on the basis of their tributary width. Thus, the lateral force acting on the shearwall at line 2 is

$$V = \sum wB$$
$$= \left(240 \frac{\text{lbf}}{\text{ft}}\right)\left(\frac{100 \text{ ft}}{2}\right) + \left(300 \frac{\text{lbf}}{\text{ft}}\right)\left(\frac{60 \text{ ft}}{2}\right)$$
$$= 21{,}000 \text{ lbf} \quad (21 \text{ kips})$$

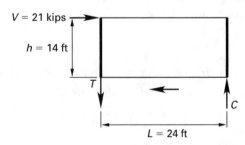

elevation of wall on line B

The overturning moment on the wall is

$$M = Vh$$
$$= (21 \text{ kips})(14 \text{ ft})$$
$$= 294 \text{ ft-kips}$$

The axial force in the shearwall boundary members is

$$T = C = \frac{M}{L}$$
$$= \frac{294 \text{ ft-kips}}{24 \text{ ft}}$$
$$= 12.3 \text{ kips} \quad (12 \text{ kips})$$

The answer is (B).

124. The height of the roof above the base is

$$h_n = 14 \text{ ft} + 12 \text{ ft}$$
$$= 26 \text{ ft}$$

The period can be approximated from the following formula.
$$T = C_t h_n^{3/4}$$
$$= (0.020)(26 \text{ ft})^{3/4}$$
$$= 0.23 \text{ sec}$$

The seismic dead load for NS ground motion includes the dead weight of second floor, roof, and the exterior walls.

$$W = (w_{D2} + w_{Dr})BL + 2w_{\text{wall}}(B+L)h$$
$$= \left(30 \frac{\text{lbf}}{\text{ft}^2} + 20 \frac{\text{lbf}}{\text{ft}^2}\right)(120 \text{ ft})(60 \text{ ft}) + (2)\left(15 \frac{\text{lbf}}{\text{ft}^2}\right)$$
$$\times \left(\frac{14 \text{ ft}}{2} + 12 \text{ ft} + 3 \text{ ft}\right)(60 \text{ ft} + 120 \text{ ft})$$
$$= 478{,}800 \text{ lbf} \quad (479 \text{ kips})$$

For a building frame system consisting of light-frame walls with wood structural shear panels, $R = 6.5$. The base shear is given as

$$V = \frac{S_{DS}}{\frac{R}{I_E}}W = \left(\frac{0.6}{\frac{6.5}{1.0}}\right)(479 \text{ kips})$$
$$= 44.2 \text{ kips}$$

The base shear need not be greater than

$$V = \frac{S_{D1}}{T\frac{R}{I_E}} = \left(\frac{0.2}{(0.23)\left(\frac{6.5}{1.0}\right)}\right)(479 \text{ kips})$$
$$= 64.0 \text{ kips}$$

In the expression above, T is the magnitude of the period and is dimensionless. The base shear must be greater than

$$V = 0.044 I_E S_{DS} W$$
$$= (0.044)(1.0)(0.6)(479 \text{ kips})$$
$$= 12.6 \text{ kips}$$

Therefore, $V = 44.2$ kips (45 kips).

The answer is (B).

125. The effective span for the slab design is the clear spacing plus one-half the flange width.

$$L_s = L_n + \frac{b_f}{2}$$

$$= (8.5 \text{ ft} - 1 \text{ ft}) + \frac{1 \text{ ft}}{2}$$

$$= 8.0 \text{ ft}$$

Based on AASHTO specifications, the impact factor cannot be greater than 0.30.

$$I \leq \frac{50 \text{ ft}}{L_s + 125 \text{ ft}}$$

$$\leq \frac{50 \text{ ft}}{8.0 \text{ ft} + 125 \text{ ft}}$$

$$\leq 0.38$$

$$I = 0.30$$

The wheel-load bending moment for an HS 20 loading is given by an AASHTO equation, with a continuity factor of 0.8.

$$M_L = (0.8)\left(\frac{L_s + 2 \text{ ft}}{32 \text{ ft}}\right)P$$

$$= (0.8)\left(\frac{8.0 \text{ ft} + 2 \text{ ft}}{32 \text{ ft}}\right)(16 \text{ kips})$$

$$= 4 \text{ ft-kips/ft}$$

The design moment for the slab is

$$M_u = 1.3\left(M_D + \frac{5}{3}(1 + I)M_L\right)$$

$$= (1.3)\left(1.0 \; \frac{\text{ft-kips}}{\text{ft}}\right.$$

$$\left. + \left(\frac{5}{3}\right)(1 + 0.30)\left(4 \; \frac{\text{ft-kips}}{\text{ft}}\right)\right)$$

$$= 12.6 \text{ ft-kips/ft} \quad (12 \text{ ft-kips/ft})$$

The answer is (D).

126. For the segment AB,

$$A_1 = \frac{\pi(D_o^2 - D_i^2)}{4} = \frac{\pi\left((3 \text{ in})^2 - (1.5 \text{ in})^2\right)}{4}$$

$$= 5.30 \text{ in}^2$$

For the segment BC,

$$A_2 = \frac{\pi D_o^2}{4} = \frac{\pi(3 \text{ in})^2}{4}$$

$$= 7.07 \text{ in}^2$$

The tension stress in segment AB is

$$f \leq F_a$$

$$\frac{P}{A} \leq F_a$$

$$\frac{P}{5.30 \text{ in}^2} \leq 22 \; \frac{\text{kips}}{\text{in}^2}$$

$$P \leq 117 \text{ kips}$$

Since the force in segment BC is smaller than in segment AB and since segment BC has a larger cross-sectional area than segment AB, BC is not the controlling section for stress. The limit on the tip deflection is

$$\Delta \leq 0.04 \text{ in}$$

$$\sum \frac{PL}{AE} \leq 0.04 \text{ in}$$

$$\frac{P(20 \text{ in})}{(5.30 \text{ in}^2)\left(29{,}000 \; \dfrac{\text{kips}}{\text{in}^2}\right)}$$

$$+ \frac{P(30 \text{ in})}{(7.07 \text{ in}^2)\left(29{,}000 \; \dfrac{\text{kips}}{\text{in}^2}\right)}$$

$$- \frac{(50 \text{ kips})(30 \text{ in})}{(7.07 \text{ in}^2)\left(29{,}000 \; \dfrac{\text{kips}}{\text{in}^2}\right)} \leq 0.04 \text{ in}$$

$$P \leq 171 \text{ kips}$$

The stress in the segment AB controls.

The answer is (B).

127. The influence line for the moment at D is obtained by cutting through point D and giving a small unit displacement so that only the unknown moment at D does internal work. The corresponding displaced shape is shown as the dashed line and is the influence line for the moment at D.

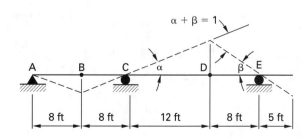

From the illustration, $\alpha + \beta = 1$. The work done by the moment through the small displacement is

$$
\begin{aligned}
W &= M_D\alpha + M_D\beta \\
&= M_D(\alpha + \beta) \\
&= M_D
\end{aligned}
$$

From geometry (small angle theory), the vertical displacement at point D is

$$
\Delta_D = (12 \text{ ft})\alpha = (8 \text{ ft})\beta
$$
$$
\alpha = \frac{2\beta}{3}
$$

Substituting,

$$
\alpha + \beta = 1
$$
$$
\frac{2\beta}{3} + \beta = 1
$$
$$
\beta = 3/5
$$

The ordinate of the influence line at point D is

$$
\begin{aligned}
\Delta_D &= (8 \text{ ft})\beta \\
&= (8 \text{ ft})\left(\frac{3}{5}\right)\left(1\,\frac{\text{kip}}{\text{kip}}\right) \\
&= 4.8 \text{ ft-kips/kip} \quad (5 \text{ ft-kips/kip})
\end{aligned}
$$

The answer is (C).

128. The influence line for shear midway between B and C, point B′, is obtained by cutting the beam and giving it a unit displacement such that only the shear at that point does any work. The resulting influence line is

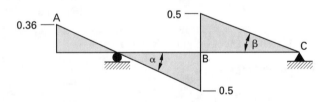

Absolute maximum shear force at B′ occurs with dead load over the full length and the uniform live load only over those regions where the ordinates of the influence line are positive.

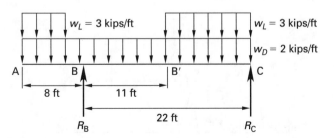

The reaction at point C is found by summing the moments about the left support. Assume clockwise moment is positive.

$$
\sum M_B = 0
$$

$$
\left(2\,\frac{\text{kips}}{\text{ft}}\right)(30 \text{ ft})(7 \text{ ft})
$$
$$
+ \left(3\,\frac{\text{kips}}{\text{ft}}\right)\big((11 \text{ ft})(22 \text{ ft} - 5.5 \text{ ft})
$$
$$
- (8 \text{ ft})(4 \text{ ft})\big) - (22 \text{ ft})R_C = 0
$$
$$
R_C = 39.5 \text{ kips}
$$
$$
(39 \text{ kips})
$$

The absolute maximum shear at B′ is

$$
\begin{aligned}
V_{B'} &= \left|-R_C + w\frac{L}{2}\right| \\
&= \left|-39.5 \text{ kips} + \left(5\,\frac{\text{kips}}{\text{ft}}\right)\left(\frac{22 \text{ ft}}{2}\right)\right| \\
&= 15.5 \text{ kips}
\end{aligned}
$$

The answer is (B).

129. Using the dummy load method, the unit virtual force is applied at D in the direction of the required deflection.

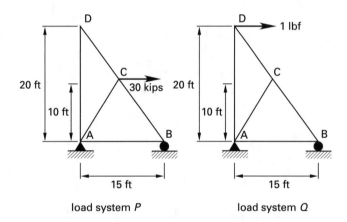

load system P load system Q

The member forces for the real loads, load system P, and for the dummy load system, system Q, are found using basic statics.

member	N_P (kips)	N_Q (lbf)	L (in)	$N_P N_Q L$ (kips-lbf-in)
AB	15.0	1.0	180	2700
AC	25.0	0	150	0
AD	0	1.33	240	0
BC	−25.0	−1.67	150	6263
CD	0	−1.67	150	0
				8963

Applying the virtual work principle,

$$(1 \text{ lbf})\Delta_{D_h} = \sum_{i=1}^{5}\left(\frac{N_P N_Q L}{AE}\right)_i$$

$$= \frac{(8963 \text{ kips-lbf-in})}{(8 \text{ in}^2)\left(29{,}000 \dfrac{\text{kips}}{\text{in}^2}\right)}$$

$$\Delta_{D_h} = 0.0386 \text{ in to the right}$$

$$(0.04 \text{ in to the right})$$

The answer is (C).

130. The rotation is obtained by applying a unit dummy couple at joint C and applying the virtual work principle.

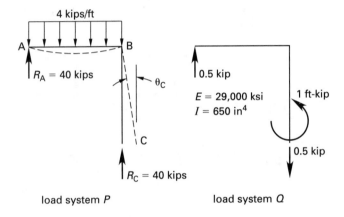

4 kips/ft

A   B

$R_A = 40$ kips θ_C

0.5 kip

$E = 29{,}000$ ksi
$I = 650$ in^4 1 ft-kip

C

0.5 kip

$R_C = 40$ kips

load system P load system Q

For load system P, the moment for member AB is

$$M_P = (40 \text{ kips})x - \left(4 \dfrac{\text{kips}}{\text{ft}}\right)\frac{x}{2}$$

$$0 \text{ ft} \le x \le 20 \text{ ft}$$

The moment for member BC is

$$M_P = 0$$

$$0 \text{ ft} \le x \le 20 \text{ ft}$$

To maintain equilibrium under the dummy loading, an upward force of 0.05 kips must act at A, with an equal and opposite force at C. Since the moment in member BC is zero throughout in load system P, the moment in member AB is needed only for load system Q.

$$M_Q = (0.05 \text{ kips})x$$

$$0 \text{ ft} \le x \le 20 \text{ ft}$$

The rotation at C is

$$(1 \text{ ft-kip})\theta_C = \int_L \frac{M_Q M_P \, dx}{EI}$$

$$= \int_{0 \text{ ft}}^{20 \text{ ft}} \frac{(0.05 \text{ kips})x \left(\begin{array}{c}(40 \text{ kips})x \\ -\left(2\dfrac{\text{kips}}{\text{ft}}\right)x^2\end{array}\right) dx}{EI}$$

$$\theta_C = \left. \frac{\left(2.0 \dfrac{\text{kip}}{\text{ft}}\right)x^3}{3EI} - \frac{\left(0.1 \dfrac{\text{kip}}{\text{ft}^2}\right)x^4}{4EI} \right|_0^{20 \text{ ft}}$$

$$= \frac{\left(\dfrac{\left(2.0\dfrac{\text{kip}}{\text{ft}}\right)(20 \text{ ft})^3}{3} - \dfrac{\left(0.1\dfrac{\text{kip}}{\text{ft}^2}\right)(20 \text{ ft})^4}{4}\right)}{\left(29{,}000 \dfrac{\text{kips}}{\text{in}^2}\right)(650 \text{ in}^4)} \times \left(144 \dfrac{\text{in}^2}{\text{ft}^2}\right)$$

$$= 0.0102 \text{ radians counterclockwise}$$

$$(0.01 \text{ radians counterclockwise})$$

The answer is (D).

131. The pile group is subjected to combined axial compression plus biaxial bending. Maximum compression occurs in the pile farthest from the pile group centroid at the location where the forces due to bending and axial compression are additive.

$$M_x = Pe_y$$
$$= (800 \text{ kips})(1.6 \text{ ft})$$
$$= 1280 \text{ ft-kips}$$
$$M_y = Pe_x$$
$$= (800 \text{ kips})(1.2 \text{ ft})$$
$$= 960 \text{ ft-kips}$$
$$J = \sum (x_i^2 + y_i^2)$$
$$= (2)(2)(4)\big((4.5 \text{ ft})^2 + (1.5 \text{ ft})^2\big) = 360 \text{ ft}^2$$
$$P_{\max} = \frac{P}{n} + \frac{M_x c_x + M_y c_y}{J}$$
$$= \frac{800 \text{ kips}}{16} + \frac{\begin{array}{c}(1280 \text{ ft-kips})(4.5 \text{ ft}) \\ + (960 \text{ ft-kips})(4.5 \text{ ft})\end{array}}{360 \text{ ft}^2}$$
$$= 78 \text{ kips} \quad (80 \text{ kips})$$

The answer is (C).

132.

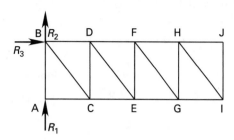

The truss has 3 external reaction components, 17 members, and 10 joints.

$$m + n_R = 2j$$
$$17 + 3 = (2)(10)$$

The necessary condition for a statically determinate truss is satisfied, but this is not sufficient to prove the truss is statically determinate. There is a problem with the support arrangement. The three reaction components are concurrent at joint B. If a nonconcurrent force is applied at any joint, it is impossible to satisfy the equilibrium condition that the summation of moments about joint B must equal zero. Therefore, the truss is unstable.

The answer is (B).

133. Release the reaction at point B to create a stable determinate beam.

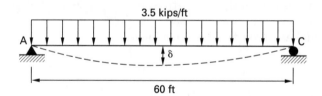

The downward displacement in the released structure at B caused by the applied uniformly distributed load is

$$\delta = \frac{5wL^4}{384EI}$$

$$= \frac{(5)\left(\left(3.5 \ \frac{\text{kips}}{\text{ft}}\right)(60 \ \text{ft})\right)\left((60 \ \text{ft})\left(12 \ \frac{\text{in}}{\text{ft}}\right)\right)^3}{(384)\left(29{,}000 \ \frac{\text{kips}}{\text{in}^2}\right)(1630 \ \text{in}^4)}$$

$$= 21.59 \ \text{in}$$

The flexibility coefficient, f, is obtained by applying a unit force upward at the released point and computing the displacement caused by that force at the released point.

$$f = \frac{L^3}{48EI}$$

$$= \frac{\left((60 \ \text{ft})\left(12 \ \frac{\text{in}}{\text{ft}}\right)\right)^3}{(48)\left(29{,}000 \ \frac{\text{kips}}{\text{in}^2}\right)(1630 \ \text{in}^4)}$$

$$= 0.1645 \ \text{in/kip}$$

For consistent displacement at point B,

$$\delta + R_B f = -0.5 \ \text{in}$$

$$-21.59 \ \text{in} + R_B\left(0.1645 \ \frac{\text{in}}{\text{kip}}\right) = -0.5 \ \text{in}$$

$$R_B = 128 \ \text{kips} \quad (130 \ \text{kips})$$

The answer is (C).

134. Due to symmetry, there are only two possible collapse mechanisms. In the first mechanism, plastic hinges form in the end spans.

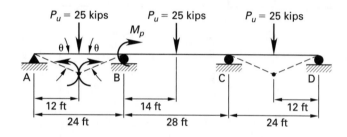

$$W_{\text{ext}} = U_{\text{int}}$$
$$P_u \delta_y = 3M_p \theta$$
$$(25 \ \text{kips})(12 \ \text{ft})\theta = 3M_p \theta$$
$$M_p = 100 \ \text{ft-kips}$$

In the second mechanism, plastic hinges form in the interior span.

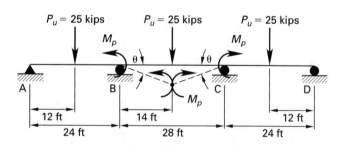

$$W_{\text{ext}} = U_{\text{int}}$$
$$P_u \delta_y = 4M_p \theta$$
$$(25 \ \text{kips})(14 \ \text{ft})\theta = 4M_p \theta$$
$$M_p = 87.5 \ \text{ft-kips}$$

Collapse is controlled by the exterior spans, so $M_p = 100$ ft-kips.

The answer is (D).

135. Let K represent the force applied at midspan producing a unit midspan deflection.

$$\frac{KL^3}{48EI} = 1$$

$$K = \frac{48EI}{L^3}$$

$$= \frac{(48)\left(29{,}000\ \frac{\text{kips}}{\text{in}^2}\right)(1600\ \text{in}^4)}{\left((16\ \text{ft})\left(12\ \frac{\text{in}}{\text{ft}}\right)\right)^3}$$

$$= 315\ \text{kips/in}$$

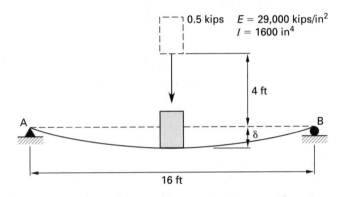

Conservation of energy requires that the potential energy of the 0.5 kips weight is converted into strain energy in the beam.

$$E_p = Wy$$
$$= (0.5\ \text{kips})(48\ \text{in} + \delta)$$
$$U_{\text{int}} = 0.5(K\delta)\delta$$
$$= (0.5)\left(315\ \frac{\text{kips}}{\text{in}}\right)\delta^2$$

Equating the two energy terms gives

$$E_p = U_{\text{int}}$$

$$(0.5\ \text{kips})(48\ \text{in} + \delta) = (0.5)\left(315\ \frac{\text{kips}}{\text{in}}\right)\delta^2$$

$$\left(315\ \frac{\text{kips}}{\text{in}}\right)\delta^2 - \delta - 48\ \text{in} = 0$$

$$\delta = 0.392\ \text{in}$$

The falling weight causes a maximum midspan deflection of 0.392 in, which corresponds to a peak force of

$$F = K\delta = \left(315\ \frac{\text{kips}}{\text{in}}\right)(0.392\ \text{in})$$

$$= 123.5\ \text{kips} \quad (120\ \text{kips})$$

The answer is (D).

136. Let K represent the force applied at the top of each column producing a unit horizontal deflection.

$$\frac{KH^3}{12EI} = 1$$

$$K = \frac{12EI}{H^3}$$

$$= \frac{(12)\left(29{,}000\ \frac{\text{kips}}{\text{in}^2}\right)(800\ \text{in}^4)}{\left((16\ \text{ft})\left(12\ \frac{\text{in}}{\text{ft}}\right)\right)^3}$$

$$= 39.3\ \text{kips/in} \quad [\text{each column}]$$

The equivalent stiffness of the four supporting columns is four times that of an individual column. The natural period of vibration of the system is

$$T = 2\pi\sqrt{\frac{m}{4K}} = 2\pi\sqrt{\frac{\dfrac{W}{g}}{4K}}$$

$$= 2\pi\sqrt{\frac{30\ \text{kips}}{\left(32.2\ \frac{\text{ft}}{\text{sec}^2}\right)(4)\left(39.3\ \frac{\text{kips}}{\text{in}}\right)\left(12\ \frac{\text{in}}{\text{ft}}\right)}}$$

$$= 0.14\ \text{sec} \quad (0.15\ \text{sec})$$

The answer is (A).

137. For $\frac{1}{2}$ in diameter 270 ksi strands, the area of one strand is $0.153\ \text{in}^2$, and the modulus of elasticity is $28{,}500\ \text{kips/in}^2$. The modulus of elasticity of the concrete at time of release is

$$E_c = 57{,}000\sqrt{f'_c}$$

$$= 57{,}000\sqrt{3500\ \frac{\text{lbf}}{\text{in}^2}}$$

$$= 3{,}370{,}000\ \text{lbf/in}^2 \quad (3370\ \text{kips/in}^2)$$

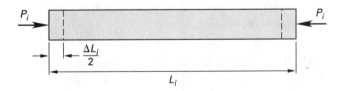

Use a trial and error method to compute loss due to elastic shortening. As a first trial, assume $\Delta f_s = 10\ \text{kips/in}^2$.

$$P_i = f_{pi}A_{ps}$$

$$= \left(200\ \frac{\text{kips}}{\text{in}^2} - 10\ \frac{\text{kips}}{\text{in}^2}\right)(4)(0.153\ \text{in}^2)$$

$$= 116\ \text{kips}$$

Following the usual assumptions for prestressed concrete, the nominal axial stress in the concrete is based on the gross concrete area.

$$f_{ci} = \frac{P_i}{A_c} = \frac{116 \text{ kips}}{(12 \text{ in})(12 \text{ in})}$$
$$= 0.806 \text{ kips/in}^2$$

For pre-tensioned members, the change in strain in the strands is the same as in the surrounding concrete. Therefore, the computed loss of prestress is the change in strain multiplied by the modulus of elasticity.

$$\Delta f_s = \frac{f_{ci}}{E_c} E_{ps}$$
$$= \left(\frac{0.806 \dfrac{\text{kips}}{\text{in}^2}}{3370 \dfrac{\text{kips}}{\text{in}^2}} \right) \left(28{,}500 \dfrac{\text{kips}}{\text{in}^2} \right)$$
$$= 6.8 \text{ kips/in}^2$$

The actual value of Δf_s is between the trial value, 10 kips/in^2, and the value computed using that trial value, 6.8 kips/in^2. For a second trial, assume 7.0 kips/in^2.

$$P_i = \left(200 \frac{\text{kips}}{\text{in}^2} - 7.0 \frac{\text{kips}}{\text{in}^2} \right) (4)(0.153 \text{ in}^2)$$
$$= 118 \text{ kips}$$
$$\Delta f_s = \frac{f_{ci}}{E_c} E_{ps}$$
$$= \left(\frac{\dfrac{118 \text{ kips}}{(12 \text{ in})(12 \text{ in})}}{3370 \dfrac{\text{kips}}{\text{in}^2}} \right) \left(28{,}500 \dfrac{\text{kips}}{\text{in}^2} \right)$$
$$= 6.9 \text{ kips/in}^2$$

The value of Δf_s is approximately 6.9 kips/in^2 (7 kips/in^2).

The answer is (B).

138. For a thin-walled pipe with internal pressure and negligible longitudinal restraint, the maximum stress occurs in the circumferential direction and can be calculated using the membrane theory of thin-walled pressure vessels.

$$f_t = \frac{pr}{t} = \frac{\left(80 \dfrac{\text{lbf}}{\text{in}^2} \right) (30 \text{ in})}{0.375 \text{ in}}$$
$$= 6400 \text{ lbf/in}^2$$

Since the longitudinal direction is unrestrained, the strain in the circumferential direction is found from the uniaxial case.

$$\epsilon = \frac{f_t}{E} = \frac{6400 \dfrac{\text{lbf}}{\text{in}^2}}{29{,}000{,}000 \dfrac{\text{lbf}}{\text{in}^2}}$$
$$= 0.00022$$

The change is diameter is directly proportional to the change in circumference.

$$\Delta D = \frac{\epsilon \pi D}{\pi}$$
$$= (0.00022)(60 \text{ in})$$
$$= 0.0132 \text{ in} \quad (0.013 \text{ in})$$

The answer is (B).

139. Relevant properties of the W12 × 106 include its area, 31.2 in^2; the moment of inertia about its weak axis, 301 in^4; the moment of inertia about its strong axis, 933 in^4; and its overall depth, 12.9 in.

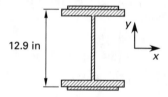

The centroid of the section is on the axes of symmetry and is located by inspection. Properties of the built-up section are

$$A = \sum A_i$$
$$= 31.2 \text{ in}^2 + (2)(10 \text{ in})(0.625 \text{ in})$$
$$= 43.7 \text{ in}^2$$
$$I_x = \sum (I_{xc} + Ad^2)$$
$$= 933 \text{ in}^4 + (2) \left(\begin{array}{c} \dfrac{(10 \text{ in})(0.625 \text{ in})^3}{12} \\ + (10 \text{ in})(0.625 \text{ in}) \\ \times \left(\dfrac{12.9 \text{ in} + 0.625 \text{ in}}{2} \right)^2 \end{array} \right)$$
$$= 1505 \text{ in}^4$$

$$I_y = \sum I_{yc}$$
$$= 301 \text{ in}^4 + (2)\left(\frac{(0.625 \text{ in})(10 \text{ in})^3}{12}\right)$$
$$= 405 \text{ in}^4$$

The major principal axis is the x-axis. The radius of gyration is

$$r_x = \sqrt{\frac{I_x}{A}}$$
$$= \sqrt{\frac{1505 \text{ in}^4}{43.7 \text{ in}^2}}$$
$$= 5.87 \text{ in}$$

The answer is (A).

140. The relevant properties of a W24 × 55 are its area, 16.3 in²; the moment of inertia about its x-axis, 1360 in⁴; and its overall depth, 23.6 in. The centroid of the built-up section is located by inspection from symmetry.

$$I_x = \sum (I_{xc} + Ad^2)$$
$$= 1360 \text{ in}^4 + (2)\left(\begin{array}{c}\frac{(8 \text{ in})(0.5 \text{ in})^3}{12} + (8 \text{ in})(0.5 \text{ in}) \\ \times \left(\frac{23.6 \text{ in} + 0.5 \text{ in}}{2}\right)^2\end{array}\right)$$
$$= 2522 \text{ in}^4$$

The shear flow between the coverplate and flange depends on the statical moment of the coverplate area about the neutral axis of bending.

$$Q = A\bar{y}$$
$$= (8 \text{ in})(0.5 \text{ in})\left(\frac{23.6 \text{ in} + 0.5 \text{ in}}{2}\right)$$
$$= 48.2 \text{ in}^3$$
$$q = \frac{VQ}{I}$$
$$= \frac{(95 \text{ kips})(48.2 \text{ in}^3)}{2522 \text{ in}^4}$$
$$= 1.8 \text{ kips/in} \quad (2 \text{ kips/in})$$

The answer is (A).

141.

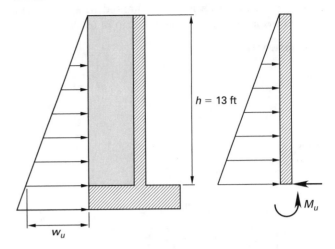

Design the stem on a per-foot-of-width basis using ACI 318. The design lateral load at the base of the stem is

$$w_u = 1.7\gamma_a h(1 \text{ ft})$$
$$= (1.7)\left(45 \frac{\text{lbf}}{\text{ft}^3}\right)(13 \text{ ft})(1 \text{ ft})$$
$$= 995 \text{ lbf/ft}$$

The design bending moment at the base is

$$M_u = \frac{0.5 w_u h^2}{3} = \frac{(0.5)\left(995 \frac{\text{lbf}}{\text{ft}}\right)(13 \text{ ft})^2}{3}$$
$$= 28{,}025 \text{ ft-lbf}$$

For a unit width of wall ($b = 12$ in) and a specified steel percentage of 0.01,

$$M_u = \phi M_n = \phi \rho f_y b d^2 \left(1 - 0.59\rho\frac{f_y}{f_c'}\right)$$

Inserting known values and equating to the value of M_u gives the required wall thickness, d.

$$(28{,}025 \text{ ft-lbf})$$
$$\times \left(12 \frac{\text{in}}{\text{ft}}\right) = (0.9)(0.01)\left(60{,}000 \frac{\text{lbf}}{\text{in}^2}\right)(12 \text{ in})d^2$$
$$\times \left(1 - (0.59)(0.01)\left(\frac{60{,}000 \frac{\text{lbf}}{\text{in}^2}}{4000 \frac{\text{lbf}}{\text{in}^2}}\right)\right)$$
$$d = 7.54 \text{ in}$$

The steel area is

$$A_s = \rho b d = (0.01)\left(12 \frac{\text{in}}{\text{ft}}\right)(7.54 \text{ in})$$
$$= 0.90 \text{ in}^2/\text{ft}$$

The answer is (D).

142. According to ACI 318, reinforced normal-weight concrete exposed to seawater is limited to a maximum water-to-cementitious-materials ratio of 0.40.

The answer is (B).

143.

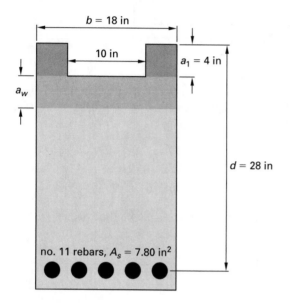

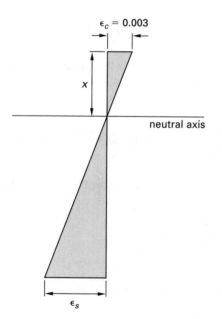

The area of concrete in compression when the section reaches its moment capacity is

$$A_c = \frac{A_s f_y}{0.85 f_c'}$$

$$= \frac{(7.80 \text{ in}^2)\left(60{,}000 \ \dfrac{\text{lbf}}{\text{in}^2}\right)}{(0.85)\left(4000 \ \dfrac{\text{lbf}}{\text{in}^2}\right)}$$

$$= 137.6 \text{ in}^2$$

This is greater than the area to the sides of the trough.

$$A_{\text{top}} = (4 \text{ in})(18 \text{ in} - 10 \text{ in})$$

$$= 32 \text{ in}^2$$

Therefore, the compression area extends below the trough to a depth of

$$a_w = \frac{A_c - A_{\text{top}}}{b}$$

$$= \frac{137.6 \text{ in}^2 - 32 \text{ in}^2}{18 \text{ in}}$$

$$= 5.87 \text{ in}$$

The depth of the neutral axis is

$$x = \frac{a}{\beta_1} = \frac{A_{\text{top}} + a_w}{\beta_1}$$

$$= \frac{4.00 \text{ in} + 5.87 \text{ in}}{0.85}$$

$$= 11.6 \text{ in}$$

From similar triangles,

$$\frac{\epsilon_s}{d - x} = \frac{0.003}{x}$$

$$\epsilon_s = (28 \text{ in} - 11.6 \text{ in})\left(\frac{0.003}{11.6 \text{ in}}\right)$$

$$= 0.004$$

The answer is (A).

144. Balanced strain conditions exist when the strain in the steel on the tension side reaches yield and, at the same time, the strain in the extreme compression edge of the concrete is at the ultimate value of 0.003.

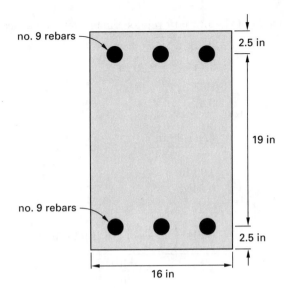

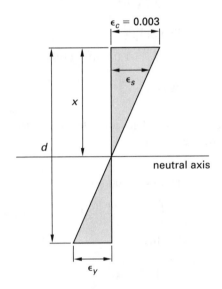

The yield strain for grade 60 rebar is

$$\epsilon_y = \frac{f_y}{E_s} = \frac{60 \ \dfrac{\text{kips}}{\text{in}^2}}{29{,}000 \ \dfrac{\text{kips}}{\text{in}^2}}$$

$$= 0.00207$$

From similar triangles,

$$\frac{x}{0.003} = \frac{d}{0.003 + \epsilon_y}$$

$$x = \frac{(21.5 \ \text{in})(0.003)}{0.003 + 0.00207}$$

$$= 12.7 \ \text{in}$$

$$\frac{\epsilon'_s}{x - 2.5 \ \text{in}} = \frac{0.003}{x}$$

$$\epsilon'_s = \frac{(12.7 \ \text{in} - 2.5 \ \text{in})(0.003)}{12.7 \ \text{in}}$$

$$= 0.0024 > \epsilon_y$$

The strain in the compression steel exceeds yield. Thus, the stress in the compression steel is the yield stress, and the force in the compression steel is

$$C'_s = f_y A'_s = \left(60 \ \frac{\text{kips}}{\text{in}^2}\right)(3.00 \ \text{in}^2)$$

$$= 180 \ \text{kips}$$

The answer is (D).

145. For a concentrically loaded tied column, the design strength is given by

$$\phi P_{n,\text{max}} = 0.8\phi\big(0.85 f'_c(A_g - A_{\text{st}}) + f_y A_{\text{st}}\big)$$

For a specified longitudinal steel ratio of 0.02,

$$A_{\text{st}} = \rho_g A_g$$

$$= 0.02 A_g$$

Substituting gives

$$1090 \ \text{kips} = \phi P_{n,\text{max}} = (0.8)(0.65)$$

$$\times \left(\begin{array}{l} (0.85)\left(5 \ \dfrac{\text{kips}}{\text{in}^2}\right)(A_g - 0.02 A_g) \\[2mm] + \left(60 \ \dfrac{\text{kips}}{\text{in}^2}\right)(0.02 A_g) \end{array}\right)$$

$$A_g = 391 \ \text{in}^2$$

$$b = \sqrt{A_g} = \sqrt{391 \ \text{in}^2}$$

$$= 19.8 \ \text{in} \quad (20 \ \text{in})$$

The answer is (C).

146. Use the appropriate reinforced concrete interaction diagram from *Design of Concrete Structures*, or the equivalent. For a column with a 20 in cross section in the direction that resists bending, the steel placement constant, γ, is found using

$$\gamma h = h - 2d'$$

$$\gamma(20 \ \text{in}) = 20 \ \text{in} - (2)(3 \ \text{in})$$

$$= 14 \ \text{in}$$

$$\gamma = 0.7$$

The reference interaction diagram requires two parameters.

$$R_n = \frac{P_u e}{\phi f'_c A_g h} = \frac{M_u}{\phi f'_c A_g h}$$

$$= \frac{(175 \ \text{ft-kips})\left(12 \ \dfrac{\text{in}}{\text{ft}}\right)}{(0.65)\left(4 \ \dfrac{\text{kips}}{\text{in}^2}\right)(20 \ \text{in})(18 \ \text{in})(20 \ \text{in})}$$

$$= 0.11$$

$$K_n = \frac{P_u}{\phi f'_c A_g} = \frac{875 \text{ kips}}{(0.65)\left(4 \dfrac{\text{kips}}{\text{in}^2}\right)(20 \text{ in})(18 \text{ in})}$$

$$= 0.93$$

From the interaction curves, interpolation at the point (0.11, 0.93) gives a longitudinal steel ratio, ρ_g, of 0.022. The required steel area is

$$\begin{aligned} A_{\text{st}} &= \rho_g A_g \\ &= (0.022)(20 \text{ in})(18 \text{ in}) \\ &= 7.92 \text{ in}^2 \quad (8 \text{ in}^2) \end{aligned}$$

The answer is (B).

147. Per ACI 318, the corbel must be designed for a tension force of at least $0.2V_u$.

$$\begin{aligned} N_{uc} &= 0.2V_u = (0.2)(66 \text{ kips}) \\ &= 13.2 \text{ kips} \end{aligned}$$

This requires a nominal steel area of

$$\begin{aligned} A_n &= \frac{N_{uc}}{\phi f_y} = \frac{13.2 \text{ kips}}{(0.75)\left(60 \dfrac{\text{kips}}{\text{in}^2}\right)} \\ &= 0.29 \text{ in}^2 \end{aligned}$$

The primary steel reinforcement must also resist a bending moment.

$$\begin{aligned} M_u &= V_u a + N_{uc}(h - d) \\ &= (66 \text{ kips})(6 \text{ in}) + (13.2 \text{ kips})(16 \text{ in} - 14 \text{ in}) \\ &= 422 \text{ in-kips} \end{aligned}$$

Taking the lever arm for the flexural couple conservatively as $0.9d$, the area of flexural steel is

$$\begin{aligned} A_f &= \frac{M_u}{\phi f_y 0.9d} \\ &= \frac{422 \text{ in-kips}}{(0.75)\left(60 \dfrac{\text{kips}}{\text{in}^2}\right)(0.9)(14 \text{ in})} \\ &= 0.74 \text{ in}^2 \end{aligned}$$

For shear friction reinforcement,

$$\begin{aligned} A_{vf} &= \frac{V_u}{\phi f_y \mu} = \frac{66 \text{ kips}}{(0.75)\left(60 \dfrac{\text{kips}}{\text{in}^2}\right)(1.4)} \\ &= 1.05 \text{ in}^2 \end{aligned}$$

The area of the primary reinforcement must be at least

$$\begin{aligned} A_s &\geq A_f + A_n \\ &\geq 0.74 \text{ in}^2 + 0.29 \text{ in}^2 \\ &\geq 1.03 \text{ in}^2 \\ A_s &\geq \frac{2A_{vf}}{3} + A_n \\ &\geq \frac{(2)(1.05 \text{ in}^2)}{3} + 0.29 \text{ in}^2 \\ &\geq 0.99 \text{ in}^2 \\ A_s &\geq \frac{0.04bd}{f_y} \\ &\geq \frac{(0.04)\left(5 \dfrac{\text{kips}}{\text{in}^2}\right)(16 \text{ in})(14 \text{ in})}{60 \dfrac{\text{kips}}{\text{in}^2}} \\ &= 0.75 \text{ in}^2 \end{aligned}$$

The controlling value is

$$A_s = 1.03 \text{ in}^2 \quad (1.05 \text{ in}^2)$$

The answer is (A).

148. The equivalent force system acting at the centroid of the footing is

$$\begin{aligned} P &= P_1 + P_2 + wbht \\ &= 400 \text{ kips} + 200 \text{ kips} \\ &\quad + \left(0.15 \dfrac{\text{kips}}{\text{ft}^3}\right)(26 \text{ ft})(6 \text{ ft})(2 \text{ ft}) \\ &= 647 \text{ kips} \\ e &= \frac{P_1 e_1 + P_2 e_2}{P} \\ &= \frac{(400 \text{ kips})(-10 \text{ ft}) + (200 \text{ kips})(12.5 \text{ ft})}{647 \text{ kips}} \\ &= -2.3 \text{ ft} \end{aligned}$$

Since the resultant force on the footing acts within the middle third of the footing's length, the entire area beneath the footing is in compression.

$$\begin{aligned} A &= bh = (6 \text{ ft})(26 \text{ ft}) \\ &= 156 \text{ ft}^2 \\ S &= \frac{hb^2}{6} = \frac{(6 \text{ ft})(26 \text{ ft})^2}{6} \\ &= 676 \text{ ft}^3 \\ f_{p,\text{max}} &= \frac{P}{A} + \frac{Pe}{S} \\ &= \frac{647 \text{ kips}}{156 \text{ ft}^2} + \frac{(647 \text{ kips})(2.3 \text{ ft})}{676 \text{ ft}^3} \\ &= 6.3 \text{ kips/ft}^2 \quad (6 \text{ kips/ft}^2) \end{aligned}$$

The answer is (D).

149.

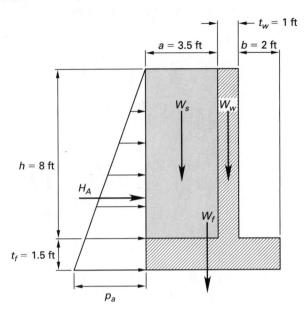

Per unit length of wall (1 ft), the force components resisting overturning are the weight of the soil, the weight of the wall stem, and the weight of the footing. The force exerted by the soil is

$$W_s = \gamma_s ah(1 \text{ ft})$$
$$= \left(100 \ \frac{\text{lbf}}{\text{ft}^3}\right)(3.5 \text{ ft})(8 \text{ ft})(1 \text{ ft})$$
$$= 2800 \text{ lbf}$$

The force exerted by the wall stem is

$$W_w = \gamma_c t_w h(1 \text{ ft})$$
$$= \left(150 \ \frac{\text{lbf}}{\text{ft}^3}\right)(1 \text{ ft})(8 \text{ ft})(1 \text{ ft})$$
$$= 1200 \text{ lbf}$$

The force exerted by the footing is

$$W_f = \gamma_c t_f(a + t_w + b)(1 \text{ ft})$$
$$= \left(150 \ \frac{\text{lbf}}{\text{ft}^3}\right)(1.5 \text{ ft})(3.5 \text{ ft} + 1 \text{ ft} + 2 \text{ ft})(1 \text{ ft})$$
$$= 1463 \text{ lbf}$$

The resisting moment about the toe of the footing is

$$M_r = \sum W_i x_i$$
$$= (2800 \text{ lbf})(4.75 \text{ ft}) + (1200 \text{ lbf})(2.5 \text{ ft})$$
$$\quad + (1463 \text{ lbf})(3.25 \text{ ft})$$
$$= 21,055 \text{ ft-lbf}$$

The active earth pressure at the base is

$$p_a = \gamma_f(h + t_f)(1 \text{ ft})$$
$$= \left(35 \ \frac{\text{lbf}}{\text{ft}^3}\right)(8 \text{ ft} + 1.5 \text{ ft})(1 \text{ ft})$$
$$= 333 \text{ lbf/ft}$$

The total active earth pressure is

$$H_A = 0.5 p_a(h + t_f)$$
$$= (0.5)\left(333 \ \frac{\text{lbf}}{\text{ft}}\right)(8 \text{ ft} + 1.5 \text{ ft})$$
$$= 1582 \text{ lbf}$$

The overturning moment is

$$M_o = H_A\left(\frac{h + t_f}{3}\right)$$
$$= (1582 \text{ lbf})\left(\frac{8 \text{ ft} + 1.5 \text{ ft}}{3}\right)$$
$$= 5010 \text{ ft-lbf}$$

The factor of safety is

$$\text{FS} = \frac{M_r}{M_o} = \frac{21,055 \text{ ft-lbf}}{5010 \text{ ft-lbf}}$$
$$= 4.2 \quad (4)$$

The answer is (D).

150. Use the customary nomenclature and methods of analysis of the *PCI Design Manual*. Replace the prestress by the statically equivalent force system acting at the centroid of concrete. The eccentricity, e, of the strands is their distance from midheight, or $(34 \text{ in}/2) - 2 \text{ in} = 15 \text{ in}$.

$$P_i = f_{pi}A_{ps} = \left(180 \ \frac{\text{kips}}{\text{in}^2}\right)(0.918 \text{ in}^2)$$
$$= 165 \text{ kips}$$
$$M_i = P_i e = (165 \text{ kips})(15 \text{ in})$$
$$= 2475 \text{ in-kips}$$
$$E_{ci} = 33 w_c^{1.5}\sqrt{f'_{ci}}$$
$$= (33)\left(110 \ \frac{\text{lbf}}{\text{ft}^3}\right)^{1.5}\sqrt{3500 \ \frac{\text{lbf}}{\text{in}^2}}$$
$$= 2{,}250{,}000 \text{ lbf/in}^2$$
$$I = \frac{bh^3}{12} = \frac{(14 \text{ in})(34 \text{ in})^3}{12}$$
$$= 45{,}854 \text{ in}^4$$
$$A = bh = (14 \text{ in})(34 \text{ in})\left(\frac{1 \text{ ft}^2}{144 \text{ in}^2}\right)$$
$$= 3.31 \text{ ft}^2$$
$$w = w_c A = \left(110 \ \frac{\text{lbf}}{\text{ft}^3}\right)(3.31 \text{ ft}^2)$$
$$= 364 \text{ lbf/ft}$$

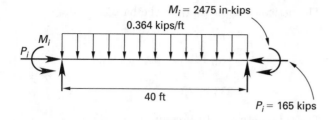

$$M_i = 2475 \text{ in-kips}$$
$$0.364 \text{ kips/ft}$$

40 ft

$$P_i = 165 \text{ kips}$$

The midspan camber is the algebraic sum of the deflections caused by the end moment and transverse beam weight.

$$\delta_i = \frac{5wL^4}{384E_{ci}I} - \frac{M_iL^2}{8E_{ci}I}$$

$$= \frac{(5)\left(0.364 \, \dfrac{\text{kips}}{\text{ft}}\right)(40 \text{ ft})^4\left(12 \, \dfrac{\text{in}}{\text{ft}}\right)^3}{(384)\left(2250 \, \dfrac{\text{kips}}{\text{in}^2}\right)(45{,}854 \text{ in}^4)}$$

$$- \frac{(2475 \text{ in-kips})(40 \text{ ft})^2\left(12 \, \dfrac{\text{in}}{\text{ft}}\right)^2}{(8)\left(2250 \, \dfrac{\text{kips}}{\text{in}^2}\right)(45{,}854 \text{ in}^4)}$$

$$= 0.203 \text{ in} - 0.691 \text{ in}$$

$$= -0.488 \text{ in} \quad (0.5 \text{ in} \uparrow)$$

The answer is (B).

---

151. Applying the usual assumptions for the analysis of prestressed beams, the tendon profile over each 40 ft segment can be represented by superposition of the chord, which is inclined upward 10 in, and by a parabolic strand that has an equivalent sag of

$$s = 10 \text{ in} + (0.5)(10 \text{ in})$$
$$= 15 \text{ in}$$

$$P = 280 \text{ kips}$$
$$P = 280 \text{ kips}$$
$$e_t = 10 \text{ in}$$

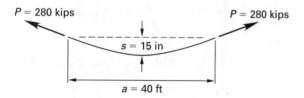

$$P = 280 \text{ kips}$$
$$P = 280 \text{ kips}$$
$$s = 15 \text{ in}$$
$$a = 40 \text{ ft}$$

The equivalent prestress forces are those of the reactions of the stressed strand against the rigid concrete.

$$w_e = \frac{8Ps}{a^2} = \frac{(8)(280 \text{ kips})\left(\dfrac{15 \text{ in}}{12 \, \dfrac{\text{in}}{\text{ft}}}\right)}{(40 \text{ ft})^2}$$

$$= 1.75 \text{ kips/ft} \uparrow$$

$$F = w_e a + \frac{2Pe_t}{a}$$

$$= \left(1.75 \, \frac{\text{kips}}{\text{ft}}\right)(40 \text{ ft}) + \frac{(2)(280 \text{ kips})(10 \text{ in})}{(40 \text{ ft})\left(12 \, \dfrac{\text{in}}{\text{ft}}\right)}$$

$$= 81.67 \text{ kips} \downarrow$$

The structure is statically indeterminate to one degree. Let the reaction at point B be the unknown force, and release it.

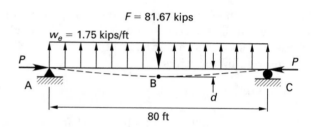

$$F = 81.67 \text{ kips}$$
$$w_e = 1.75 \text{ kips/ft}$$
$$P \qquad\qquad P$$
A   B d C
80 ft

$$EI = 250 \times 10^6 \text{ kips-in}^2$$

$$d = \frac{5w_eL^4}{384EI} - \frac{FL^3}{48EI}$$

$$= \frac{(5)\left(1.75 \, \dfrac{\text{kips}}{\text{ft}}\right)(80 \text{ ft})^4\left(12 \, \dfrac{\text{in}}{\text{ft}}\right)^3}{(384)(250 \times 10^6 \text{ kips-in}^2)}$$

$$- \frac{(81.67 \text{ kips})(80 \text{ ft})^3\left(12 \, \dfrac{\text{in}}{\text{ft}}\right)^3}{(48)(250 \times 10^6 \text{ kips-in}^2)}$$

$$= 0.430 \text{ in} \uparrow$$

The flexibility coefficient is obtained by applying a unit force upward at the released point and computing the deflection caused by that force.

$$f = \frac{L^3}{48EI} = \frac{(80 \text{ ft})^3\left(12 \, \dfrac{\text{in}}{\text{ft}}\right)^3}{(48)(250 \times 10^6 \text{ kips-in}^2)}$$

$$= 0.0737 \text{ in/kip} \uparrow$$

For consistent displacement,

$$d + R_B f = 0$$

$$0.430 \text{ in} + R_B \left(0.0737 \, \frac{\text{in}}{\text{kip}} \right) = 0$$

$$R_B = 5.8 \text{ kips} \downarrow$$

$$(6 \text{ kips} \downarrow)$$

The answer is (B).

152.

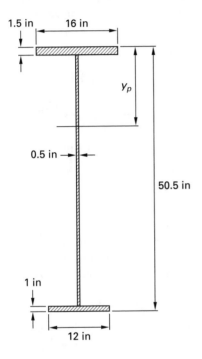

The section consists of three rectangles having areas A_1, A_2, and A_3.

$$A_i = b_i h_i$$
$$A_1 = (16 \text{ in})(1.5 \text{ in})$$
$$= 24 \text{ in}^2$$
$$A_2 = (0.5 \text{ in})(48 \text{ in})$$
$$= 24 \text{ in}^2$$
$$A_3 = (12 \text{ in})(1 \text{ in})$$
$$= 12 \text{ in}^2$$

For the fully plastic condition, the plastic neutral axis is positioned such that the area in compression equals the area in tension.

$$A_{\text{comp}} = A_{\text{ten}} = \frac{\sum A_i}{2}$$
$$= \frac{24 \text{ in}^2 + 24 \text{ in}^2 + 12 \text{ in}^2}{2}$$
$$= 30 \text{ in}^2$$

$$y_p = 1.5 \text{ in} + \frac{A_{\text{comp}} - A_1}{b_2}$$
$$= 1.5 \text{ in} + \frac{30 \text{ in}^2 - 24 \text{ in}^2}{0.5 \text{ in}}$$
$$= 13.5 \text{ in}$$

The region above the plastic neutral axis is under uniform compression stress, F_y, and the region below is under uniform tension stress, which is also F_y. Taking moments of the stress resultants gives the plastic moment capacity. For convenience, take the moments about the tension force in the bottom flange

$$M_p = \sum F_y A_i \bar{y}_i$$
$$= \left(36 \, \frac{\text{kips}}{\text{in}^2} \right) \left(\begin{array}{c} (16 \text{ in})(1.5 \text{ in})(50 \text{ in} - 0.75 \text{ in}) \\ + (0.5 \text{ in})(12 \text{ in})(50 \text{ in} - 7.5 \text{ in}) \\ - (0.5 \text{ in})(36 \text{ in})(18.5 \text{ in}) \end{array} \right)$$
$$\times \left(\frac{1 \text{ ft}}{12 \text{ in}} \right)$$
$$= 3312 \text{ ft-kips} \quad (3300 \text{ ft-kips})$$

The answer is (C).

153. For the ASD option, since the moment is zero at the support, the ratio M_1/M_2 is zero. For this case,

$$C_b = 1.75 + 1.05 \, \frac{M_1}{M_2} + 0.3 \left(\frac{M_1}{M_2} \right)^2$$
$$= 1.75 \quad (1.8)$$

The answer is (D).

For the LRFD option, the design load is controlled by

$$w_u = 1.2 w_D + 1.6 w_L$$
$$= (1.2) \left(2.5 \, \frac{\text{kips}}{\text{ft}} \right) + (1.6) \left(1.8 \, \frac{\text{kips}}{\text{ft}} \right)$$
$$= 5.88 \text{ kips/ft}$$

For a uniformly loaded simple beam,

$$M_{u,x} = \frac{w_u L x}{2} - \frac{w_u x^2}{2}$$
$$M_{\text{max}} = \frac{w_u L^2}{8} = \frac{\left(5.88 \, \frac{\text{kips}}{\text{ft}} \right) (36 \text{ ft})^2}{8}$$
$$= 953 \text{ ft-kips}$$

$$M_A = \frac{\left(5.88 \; \frac{\text{kips}}{\text{ft}}\right)(36 \text{ ft})(4.5 \text{ ft})}{2}$$
$$- \frac{\left(5.88 \; \frac{\text{kips}}{\text{ft}}\right)(4.5 \text{ ft})^2}{2}$$
$$= 417 \text{ ft-kips}$$

$$M_B = \frac{\left(5.88 \; \frac{\text{kips}}{\text{ft}}\right)(36 \text{ ft})(9 \text{ ft})}{2}$$
$$- \frac{\left(5.88 \; \frac{\text{kips}}{\text{ft}}\right)(9 \text{ ft})^2}{2}$$
$$= 714 \text{ ft-kips}$$

$$M_C = \frac{\left(5.88 \; \frac{\text{kips}}{\text{ft}}\right)(36 \text{ ft})(13.5 \text{ ft})}{2}$$
$$- \frac{\left(5.88 \; \frac{\text{kips}}{\text{ft}}\right)(13.5 \text{ ft})^2}{2}$$
$$= 893 \text{ ft-kips}$$

The bending coefficient is

$$C_b = \frac{12.5 M_{\text{max}}}{2.5 M_{\text{max}} + 3 M_A + 4 M_B + 3 M_C}$$

$$= \frac{(12.5)(953 \text{ ft-kips})}{(2.5)(953 \text{ ft-kips}) + (3)(417 \text{ ft-kips})}$$
$$+ (4)(714 \text{ ft-kips}) + (3)(893 \text{ ft-kips})$$

$$= 1.30 \quad (1.3)$$

The answer is (D).

154. For the ASD option, the strength of the two fillet welds based on their effective throat area is

$$f_w = (2)(0.707w)0.3F_u$$

$$= (2)(0.707)(0.25 \text{ in})(0.3) \left(70 \; \frac{\text{kips}}{\text{in}^2}\right)$$

$$= 7.42 \text{ kips/in}$$

To avoid overstressing the base material, the AISC ASD specification limits the shear stress in the base material adjacent to the welds to $0.4F_y$. The thickness must be

$$t \geq \frac{f_w}{0.4 F_y} \geq \frac{7.42 \; \frac{\text{kips}}{\text{in}}}{(0.4)\left(36 \; \frac{\text{kips}}{\text{in}^2}\right)}$$

$$\geq 0.52 \text{ in} \quad (0.5 \text{ in})$$

The answer is (C).

For the LRFD option, the strength of the two fillet welds based on their effective throat area is

$$\phi f_w = \phi 0.707 w 0.6 F_u$$

$$= (0.75)(0.707)(2)(0.25 \text{ in})(0.6) \left(70 \; \frac{\text{kips}}{\text{in}^2}\right)$$

$$= 11.1 \text{ kips/in}$$

To avoid overloading the base material, the AISC LRFD specification requires that the shear rupture strength of the base material be greater than or equal to the weld's design strength. The thickness must be

$$t \geq \frac{\phi f_w}{\phi 0.6 F_u}$$

$$\geq \frac{11.1 \; \frac{\text{kips}}{\text{in}}}{(0.75)(0.6)\left(58 \; \frac{\text{kips}}{\text{in}^2}\right)}$$

$$\geq 0.43 \text{ in} \quad (0.4 \text{ in})$$

The answer is (C).

155. For the ASD option, the net area is the gross area less the area of four flange holes and two web holes. Since holes are punched, the hole diameter is taken as $1/8$ in greater than the fastener diameter.

$$A_n = A - 4 t_f D - 2 t_w D$$
$$= 14.4 \text{ in}^2 - (4)(0.56 \text{ in})(0.875 \text{ in})$$
$$- (2)(0.34 \text{ in})(0.875 \text{ in})$$
$$= 11.85 \text{ in}^2$$

Since all elements are connected, the shear lag coefficient, U, is 1.0.

$$A_e = U A_n$$
$$= (1.0)(11.85 \text{ in}^2)$$
$$= 11.85 \text{ in}^2$$

The allowable axial tension is

$$P \leq 0.6 F_y A_g$$
$$\leq (0.6)\left(50 \; \frac{\text{kips}}{\text{in}^2}\right)(14.4 \text{ in}^2)$$
$$= 432 \text{ kips}$$
$$P \leq 0.5 F_u A_e$$
$$= (0.5)\left(65 \; \frac{\text{kips}}{\text{in}^2}\right)(11.85 \text{ in}^2)$$
$$= 385 \text{ kips}$$

The controlling value is

$$P = 385 \text{ kips}$$

The answer is (B).

For the LRFD option, the net area is the gross area less the area of four flange holes and two web holes. Since holes are punched, the hole diameter is taken as $\frac{1}{8}$ in greater than the fastener diameter.

$$A_n = A - 4t_f D - 2t_w D$$

$$= 14.4 \text{ in}^2 - (4)(0.56 \text{ in})(0.875 \text{ in})$$

$$- (2)(0.34 \text{ in})(0.875 \text{ in})$$

$$= 11.85 \text{ in}^2$$

Since all elements are connected, the shear lag coefficient, U, is 1.0. The design axial tension strength is

$$\phi P_n \le \phi F_y A_g$$

$$\le (0.9) \left(50 \, \frac{\text{kips}}{\text{in}^2} \right) (14.4 \text{ in}^2)$$

$$= 648 \text{ kips}$$

$$\phi P_n \le \phi F_u A_e$$

$$\le (0.75) \left(65 \, \frac{\text{kips}}{\text{in}^2} \right) (11.85 \text{ in}^2)$$

$$\le 578 \text{ kips}$$

The controlling value is

$$\phi P_n = 578 \text{ kips}$$

The answer is (B).

156. The properties of a C12 × 30 are $A = 8.82 \text{ in}^2$, $t_w = 0.51$ in, $I_x = 162 \text{ in}^4$ (strong axis), $I_y = 5.12 \text{ in}^4$ (weak axis), and the centroid is located 0.674 in from the outside edge of the web. For the built-up section,

$$I_x = \sum (I_{xc} + Ad^2)$$

$$= (2) \left(\begin{array}{c} \dfrac{(0.5 \text{ in})(11 \text{ in})^3}{12} + 5.12 \text{ in}^4 \\[2mm] + (8.82 \text{ in}^2)(6 \text{ in} - 0.674 \text{ in})^2 \end{array} \right)$$

$$= 621.5 \text{ in}^4 \quad (622 \text{ in}^4)$$

$$I_y = \sum (I_{yc} + Ad^2)$$

$$= (2) \left(\begin{array}{c} \dfrac{(11 \text{ in})(0.5 \text{ in})^3}{12} \\[2mm] + (11 \text{ in})(0.5 \text{ in})(6.25 \text{ in})^2 \\[2mm] + 162 \text{ in}^4 \end{array} \right)$$

$$= 753.9 \text{ in}^4 \quad (754 \text{ in}^4)$$

$$A = \sum A_i$$

$$= (2) \big((11 \text{ in})(0.5 \text{ in}) + 8.82 \text{ in}^2 \big)$$

$$= 28.6 \text{ in}^2$$

$$r_x = \sqrt{\frac{I_x}{A}} = \sqrt{\frac{622 \text{ in}^4}{28.6 \text{ in}^2}}$$

$$= 4.66 \text{ in}$$

$$r_y = \sqrt{\frac{I_y}{A}} = \sqrt{\frac{754 \text{ in}^4}{28.6 \text{ in}^2}}$$

$$= 5.13 \text{ in}$$

The radius of gyration about the x-axis controls.

$$\frac{KL}{r} = \frac{KL_x}{r_x}$$

$$= \frac{(16 \text{ ft}) \left(12 \, \dfrac{\text{in}}{\text{ft}} \right)}{4.66 \text{ in}}$$

$$= 41$$

For the ASD option,

$$C_c = \sqrt{\frac{2\pi^2 E}{F_y}}$$

$$= \sqrt{\frac{2\pi^2 \left(29{,}000 \, \dfrac{\text{kips}}{\text{in}^2} \right)}{36 \, \dfrac{\text{kips}}{\text{in}^2}}}$$

$$= 126 > \frac{KL}{r}$$

Use the appropriate AISC equation.

$$F_a = \frac{\left(1 - \dfrac{\left(\dfrac{KL}{r}\right)^2}{2C_c^2} \right) F_y}{\dfrac{5}{3} + \dfrac{3\dfrac{KL}{r}}{8C_c} + \dfrac{\left(\dfrac{KL}{r}\right)^3}{8C_c^3}}$$

$$= \frac{\left(1 - \dfrac{(41)^2}{(2)(126)^2} \right) \left(36 \, \dfrac{\text{kips}}{\text{in}^2} \right)}{\dfrac{5}{3} + \dfrac{(3)(41)}{(8)(126)} + \dfrac{(41)^3}{(8)(126)^3}}$$

$$= 19.0 \text{ kips/in}^2$$

$$P = F_a A$$

$$= \left(19.0 \, \frac{\text{kips}}{\text{in}^2} \right) (28.6 \text{ in}^2)$$

$$= 543 \text{ kips} \quad (550 \text{ kips})$$

The answer is (A).

For the LRFD option,

$$\lambda_c = \frac{KL}{r\pi}\sqrt{\frac{F_y}{E}} = \frac{41}{\pi}\sqrt{\frac{36\ \dfrac{\text{kips}}{\text{in}^2}}{29{,}000\ \dfrac{\text{kips}}{\text{in}^2}}}$$

$$= 0.46 < 1.5$$

Use the appropriate AISC equation.

$$F_{cr} = (0.658^{\lambda_c^2})F_y$$

$$= (0.658^{(0.46)^2})\left(36\ \frac{\text{kips}}{\text{in}^2}\right)$$

$$= 32.95\ \text{kips/in}^2$$

$$\phi P_n = \phi_c F_{cr} A_g$$

$$= (0.85)\left(32.95\ \frac{\text{kips}}{\text{in}^2}\right)(28.6\ \text{in}^2)$$

$$= 801\ \text{kips}$$

The answer is (A).

157. For the 5.125 in × 22.5 in section,

$$S = \frac{bh^2}{6} = \frac{(5.125\ \text{in})(22.5\ \text{in})^2}{6}$$

$$= 432\ \text{in}^3$$

$$M_r = F_b'S$$

The allowable bending stress is obtained by multiplying the basic stress by the lesser of the volume factor, C_v, and the lateral stability factor, C_L.

$$C_v = K_L\left(\left(\frac{21}{L}\right)\left(\frac{12}{d}\right)\left(\frac{5.125}{b}\right)\right)^{1/x}$$

$$= (1.0)\left(\left(\frac{21\ \text{ft}}{36\ \text{ft}}\right)\left(\frac{12\ \text{in}}{22.5\ \text{in}}\right)\left(\frac{5.125\ \text{in}}{5.125\ \text{in}}\right)\right)^{1/10}$$

$$= 0.89$$

$$\frac{L_u}{d} = \frac{(18\ \text{ft})\left(12\ \dfrac{\text{in}}{\text{ft}}\right)}{22.5\ \text{in}}$$

$$= 9.6$$

$$7 < \frac{L_u}{d} < 14.3$$

Therefore,

$$L_e = 1.63L_u + 3d$$

$$= (1.63)(18\ \text{ft})\left(12\ \frac{\text{in}}{\text{ft}}\right) + (3)(22.5\ \text{in})$$

$$= 419.6\ \text{in}\quad (420\ \text{in})$$

$$R_B = \sqrt{\frac{L_e d}{b^2}} = \sqrt{\frac{(420\ \text{in})(22.5\ \text{in})}{(5.125\ \text{in})^2}}$$

$$= 19$$

$$F_b^* = F_b = 2400\ \text{lbf/in}^2$$

$$F_{bE} = \frac{K_{bE}E'}{R_B^2} = \frac{(0.439)\left(1{,}600{,}000\ \dfrac{\text{lbf}}{\text{in}^2}\right)}{(19)^2}$$

$$= 1946\ \text{lbf/in}^2$$

$$\frac{F_{bE}}{F_b^*} = \frac{1946\ \dfrac{\text{lbf}}{\text{in}^2}}{2400\ \dfrac{\text{lbf}}{\text{in}^2}} = 0.81$$

$$C_L = \frac{1 + \dfrac{F_{be}}{F_b^*}}{1.9} - \sqrt{\left(\frac{1 + \dfrac{F_{be}}{F_b^*}}{1.9}\right)^2 - \dfrac{\dfrac{F_{be}}{F_b^*}}{1.9}}$$

$$= \frac{1 + 0.81}{1.9} - \sqrt{\left(\frac{1 + 0.81}{1.9}\right)^2 - \frac{0.81}{0.95}}$$

$$= 0.72$$

Allowable bending stress is controlled by the stability factor, C_L. Therefore,

$$M_r = F_b'S = C_L F_b S$$

$$= \frac{(0.72)\left(2400\ \dfrac{\text{lbf}}{\text{in}^2}\right)(432\ \text{in}^3)}{\left(1000\ \dfrac{\text{lbf}}{\text{kip}}\right)\left(12\ \dfrac{\text{in}}{\text{ft}}\right)}$$

$$= 62\ \text{ft-kips}$$

The answer is (A).

158.

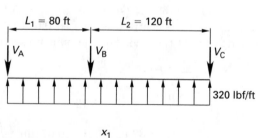

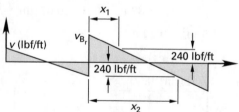

For a plywood diaphragm, the reactions to the shearwalls are in proportion to their tributary lengths.

The shear in the diaphragm just to the right of shearwall B is

$$V_{B_r} = \frac{wL_2}{2} = \frac{\left(320\ \frac{\text{lbf}}{\text{ft}}\right)(120\ \text{ft})}{2}$$
$$= 19{,}200\ \text{lbf}$$

The shear distributes uniformly over the width of the diaphragm.

$$v_{B_r} = \frac{V_{B_r}}{B} = \frac{19{,}200\ \text{lbf}}{40\ \text{ft}}$$
$$= 480\ \text{lbf/ft}$$

The variation in diaphragm shear between B and C is

$$v = v_{B_r} - mx$$
$$= 480\ \frac{\text{lbf}}{\text{ft}} - \left(\frac{960\ \frac{\text{lbf}}{\text{ft}}}{120\ \text{ft}}\right)x$$

The limits of the region where the absolute value of v is less than 240 lbf/ft are given by the following equation.

$$v = 240\ \frac{\text{lbf}}{\text{ft}} = \left|480\ \frac{\text{lbf}}{\text{ft}} - \left(8\ \frac{\text{lbf}}{\text{ft}^2}\right)x\right|$$
$$x_1 = 30\ \text{ft}$$
$$x_2 = 90\ \text{ft}$$

The answer is (C).

159. Design on a unit-length basis ($b = 12$ in).

$$M_{\max} = \frac{wH^2}{8} = \frac{\left(20\ \frac{\text{lbf}}{\text{ft}}\right)(15.33\ \text{ft})^2}{8}$$
$$= 587.5\ \text{ft-lbf} \quad (588\ \text{ft-lbf})$$

For wall reinforcement centered in a nominal 8 in masonry wall,

$$d = \frac{t}{2} = \frac{7.62\ \text{in}}{2}$$
$$= 3.81\ \text{in}$$

For grade 60 reinforcement subject to wind load, a one-third increase in allowable stress is permitted.

$$F_s = \left(\frac{4}{3}\right)\left(24{,}000\ \frac{\text{lbf}}{\text{in}^2}\right)$$
$$= 32{,}000\ \text{lbf/in}^2$$

For the masonry,

$$F_c = \left(\frac{4}{3}\right)\left(\frac{f'_m}{3}\right) = \left(\frac{4}{3}\right)\left(\frac{1500\ \frac{\text{lbf}}{\text{in}^2}}{3}\right)$$
$$= 667\ \text{lbf/in}^2$$

A trial value of A_s is found using an internal lever arm of $^7/_8$ of the effective depth.

$$A_s \approx \frac{M_{\max}}{F_s \frac{7}{8}d}$$
$$= \frac{(588\ \text{ft-lbf})\left(12\ \frac{\text{in}}{\text{ft}}\right)}{\left(32{,}000\ \frac{\text{lbf}}{\text{in}^2}\right)\left(\frac{7}{8}\right)(3.81\ \text{in})}$$
$$= 0.066\ \text{in}^2 \quad (0.07\ \text{in}^2)$$

The steel ratio is

$$\rho = \frac{A_s}{bd} = \frac{0.07\ \text{in}^2}{(12\ \text{in})(3.81\ \text{in})}$$
$$= 0.00153$$

The modular ratio is

$$n = \frac{E_s}{E_m} = \frac{E_s}{750 f'_m}$$
$$= \frac{29{,}000{,}000\ \frac{\text{lbf}}{\text{in}^2}}{(750)\left(1500\ \frac{\text{lbf}}{\text{in}^2}\right)}$$
$$= 25.77 \quad (26)$$
$$\rho n = (0.00153)(26)$$
$$= 0.04$$

The depth to the neutral axis is

$$k = \sqrt{(\rho n)^2 + 2\rho n} - \rho n$$
$$= \sqrt{(0.04)^2 + (2)(0.04)} - 0.04$$
$$= 0.246$$
$$j = 1 - \frac{k}{3}$$
$$= 1 - \frac{0.246}{3}$$
$$= 0.918$$

The stress in the steel is

$$f_s = \frac{M_{\max}}{A_s j d}$$
$$= \frac{(588\ \text{ft-lbf})\left(12\ \frac{\text{in}}{\text{ft}}\right)}{(0.07\ \text{in}^2)(0.918)(3.81\ \text{in})}$$
$$= 28{,}820\ \frac{\text{lbf}}{\text{in}^2} < F_s$$

The stress in concrete is

$$
\begin{aligned}
f_c &= \left(\frac{M_{\text{max}}}{bd^2}\right)\left(\frac{2}{jk}\right) \\
&= \left(\frac{(588 \text{ ft-lbf})\left(12 \ \frac{\text{in}}{\text{ft}}\right)}{(12 \text{ in})(3.81 \text{ in})^2}\right)\left(\frac{2}{(0.918)(0.246)}\right) \\
&= 359 \ \frac{\text{lbf}}{\text{in}^2} < F_c
\end{aligned}
$$

The trial value of $A_s = 0.07 \text{ in}^2$ is sufficient.

The answer is (B).

160. The number of loading cycles is

$$
\begin{aligned}
N &= \left(50 \ \frac{\text{cycles}}{\text{day}}\right)\left(365 \ \frac{\text{days}}{\text{yr}}\right)(25 \text{ yr design life}) \\
&= 456{,}250 \text{ cycles}
\end{aligned}
$$

Per ASD specification, the number of cycles corresponds to loading condition 2. The bottom flange will experience alternating tension stress adjacent to the transverse weld, which corresponds to stress category E'. (The flange thickness of the W21 × 93 is 0.93 in, which exceeds the limit of 0.8 in that distinguishes categories E and E'.) For this condition, ASD limits the stress range to 9 kips/in^2.

The answer is (A).

Solutions
Transportation

161. The curve radius is

$$R = \frac{(360°)(100 \text{ ft})}{2\pi D} = \frac{(360°)(100 \text{ ft})}{2\pi(2°)}$$
$$= 2864.789 \text{ ft}$$

The interior angle is

$$I = 54°56'24'' - 32°15'18''$$
$$= 22°41'6'' \quad (22.685°)$$

The tangent length between PI and PT is

$$T = R\tan\frac{I}{2} = (2864.789 \text{ ft})\tan\left(\frac{22.685°}{2}\right)$$
$$= 574.652 \text{ ft}$$

The northing (i.e., change in north dimension) is

$$\Delta N = T\cos(\text{bearing})$$
$$= (574.652 \text{ ft})\cos(54°56'24'')$$
$$= 330.100 \text{ ft}$$

The easting (i.e., change in east dimension) is

$$\Delta E = T\sin(\text{bearing})$$
$$= (574.652 \text{ ft})\sin(54°56'24'')$$
$$= 470.382 \text{ ft}$$

The northing of PT is

$$N_{\text{PT}} = N_{\text{PI}} + \Delta N$$
$$= 423,968.68 \text{ ft} + 330.10 \text{ ft}$$
$$= 424,298.78 \text{ ft}$$

The easting of the PT is

$$E_{\text{PT}} = E_{\text{PI}} + \Delta E$$
$$= 268,236.42 \text{ ft} + 470.38 \text{ ft}$$
$$= 268,706.80 \text{ ft}$$

The coordinates of the PT are 424,298.78 N and 268,706.80 E.

The answer is (A).

162. The bearing of the radius, B_r, at the PT is offset 90° from the tangent at the PT.

$$B_r = 90° + B_t$$
$$= 90° + \text{N } 54°56'24'' \text{ E}$$
$$= \text{N } 144°56'24'' \text{ E} \quad (\text{S } 35°3'36'' \text{ E})$$

The answer is (D).

163. Use the Pythagorean theorem and the given coordinates to calculate the required distance.

$$D = \sqrt{(N_{\text{cemetery}} - N_{\text{curve}})^2 + (E_{\text{cemetery}} - E_{\text{curve}})^2}$$
$$= \sqrt{\begin{array}{l}(424,239.72 \text{ ft} - 424,180.59 \text{ ft})^2 \\ \quad + (268,498.69 \text{ ft} - 268,549.70 \text{ ft})^2\end{array}}$$
$$= 78.09 \text{ ft} \quad (78.1 \text{ ft})$$

The answer is (C).

164. Use the AASHTO Green Book. For rural highways, assume the maximum rate of superelevation, 0.08-0.10. For 40 mph design speed and a 2° curve, the curve radius is

$$R = \frac{(360°)(100 \text{ ft})}{2\pi D} = \frac{(360°)(100 \text{ ft})}{2\pi(2°)}$$
$$= 2864.79 \text{ ft}$$

In the Green Book exhibits, the resulting rate of superelevation is approximately 0.03.

The answer is (B).

165. Use the AASHTO Green Book exhibit on stopping sight distance. The minimum stopping sight distance for a design speed of 40 mph is 300.6 ft (rounded up to 310 ft).

The answer is (C).

166. The curve length is the distance from the PC to the PT.

$$L = \frac{2\pi R I}{360°} = \frac{2\pi (2080 \text{ ft})(60°)}{360°}$$
$$= 2178.17 \text{ ft}$$
$$\text{sta PT} = \text{sta PC} + L = (\text{sta } 12 + 40) + 2178.17 \text{ ft}$$
$$= \text{sta } 34 + 18.17 \quad (\text{sta } 34 + 18)$$

The answer is (C).

167. From the given table, the runoff for a curve of 2080 ft radius is 150 ft. On circular curves such as this, the 150 ft represents one third of the superelevation runoff, L (the transition from normal cross slope to the superelevated cross section). Therefore, two thirds of the superelevation runoff is developed on the tangent, and in this case, that is 300 ft. The tangent runout, T_R (the distance until the adverse cross slope has been removed), is 300 ft (twice the runoff in the curve according to the problem statement). The station of the transition start (TS) is

$$\text{sta TS} = \text{sta PC} - \tfrac{2}{3}L - T_R$$
$$= (\text{sta } 12 + 40) - 300 \text{ ft} - 300 \text{ ft}$$
$$= \text{sta } 6 + 40$$

The answer is (A).

168. The midpoint, M, of the curve is located half of the curve length L past the PC.

$$\text{sta M} = \text{sta PC} + \frac{L}{2}$$
$$= (\text{sta } 13 + 50) + \frac{2000 \text{ ft}}{2}$$
$$= \text{sta } 23 + 50$$

The elevation of the centerline at M is

$$\text{elev}_{M,\text{centerline}} = \text{elev}_{PC} + G\frac{L}{2}$$
$$= 170 \text{ ft} + \left(0.0075 \ \frac{\text{ft}}{\text{ft}}\right)\left(\frac{2000 \text{ ft}}{2}\right)$$
$$= 177.5 \text{ ft}$$

Since the midpoint is more than 150 ft from the PC and more than 150 ft from the PT, the section is fully elevated at the midpoint. The cross slope is 0.045 ft/ft. The outside pavement edge is higher than the centerline by

$$\Delta_{\text{elev}} = \left(0.045 \ \frac{\text{ft}}{\text{ft}}\right)\left(12 \ \frac{\text{ft}}{\text{lane}}\right)(1 \text{ lane})$$
$$= 0.54 \text{ ft}$$

The elevation of the outside pavement edge is

$$\text{elev}_{M,\text{edge}} = \text{elev}_{M,\text{centerline}} + \Delta_{\text{elev}}$$
$$= 177.5 \text{ ft} + 0.54 \text{ ft}$$
$$= 178.04 \text{ ft} \quad (178 \text{ ft})$$

The answer is (D).

169. The peak hour factor (PHF) is defined in the *Highway Capacity Manual* (HCM) as the ratio of the total hourly volume to the peak rate of flow within the hour.

$$\text{PHF} = \frac{\text{hourly volume}}{\text{peak rate of flow within the hour}} = \frac{V}{4V_{15}}$$
$$= \frac{195 \text{ veh} + 163 \text{ veh} + 157 \text{ veh} + 178 \text{ veh}}{(4)(195 \text{ veh})}$$
$$= 0.888 \quad (0.89)$$

The answer is (B).

170. The recreational vehicles can be combined with trucks because the percentage of trucks and buses (22%) is at least five times the percentage of RVs (4%). Therefore, the analysis will be based on a percentage of trucks, P_T, of 26%.

From the HCM, the passenger car equivalent of heavy trucks, E_T, is 1.5. The heavy vehicle adjustment factor, f_{HV}, is determined using the corresponding HCM equation,

$$f_{HV} = \frac{1}{1 + P_T(E_T - 1)}$$
$$= \frac{1}{1 + (0.26)(1.5 - 1)}$$
$$= 0.885$$

The 15 min passenger car equivalent flow rate is determined from a HCM equation. The driver population factor, f_p, is assumed to be 1.0 for commuter traffic.

$$v_p = \frac{V}{(\text{PHF})Nf_{HV}f_p}$$
$$= \frac{2500 \ \frac{\text{veh}}{\text{hr}}}{(0.85)(3 \text{ lanes})(0.885)(1.0)}$$
$$= 1108 \text{ pcphpl} \quad (1110 \text{ pcphpl})$$

The answer is (B).

171. The actual free-flow speed (FFS) can be determined from a HCM equation.

$$BFFS = 65 \text{ mph} \quad [\text{given}]$$
$$f_{LW} = 1.9 \text{ mph} \quad [\text{HCM}]$$
$$f_{LC} = 0.80 \text{ mph} \quad [\text{HCM}]$$
$$f_N = 1.5 \text{ mph} \quad [\text{HCM}]$$
$$f_{ID} = 0.0 \text{ mph} \quad [\text{HCM}]$$
$$FFS = BFFS - f_{LW} - f_{LC} - f_N - f_{ID}$$
$$= 65 \text{ mph} - 1.9 \text{ mph} - 0.8 \text{ mph}$$
$$- 1.5 \text{ mph} - 0.0 \text{ mph}$$
$$= 60.8 \text{ mph} \quad (61 \text{ mph})$$

The FFS is used as the average passenger car speed, S. The density of flow from the HCM equation is

$$D = \frac{v_p}{S} = \frac{1667 \text{ pcphpl}}{61 \dfrac{\text{mi}}{\text{hr}}}$$
$$= 27.3 \text{ pcpmpl}$$

From the relevant HCM LOS exhibit, the LOS is D.

The answer is (C).

172. Assuming three lanes in each direction, the free-flow speed (FFS) can be determined from the HCM equation.

$$BFFS = 70 \text{ mph} \quad [\text{given}]$$
$$f_{LW} = 0.0 \text{ mph} \quad [\text{HCM, assuming ideal 12 ft lanes}]$$
$$f_{LC} = 0.0 \text{ mph} \quad \left[\begin{array}{c}\text{HCM, assuming ideal} \\ \text{lateral clearance} \geq 6 \text{ ft}\end{array}\right]$$
$$f_N = 3.0 \text{ mph} \quad \left[\begin{array}{c}\text{HCM, assuming three lanes} \\ \text{in each direction}\end{array}\right]$$
$$f_{ID} = 2.5 \text{ mph} \quad [\text{HCM}]$$
$$FFS = BFFS - f_{LW} - f_{LC} - f_N - f_{ID}$$
$$= 70 \text{ mph} - 0.0 \text{ mph} - 0.0 \text{ mph} - 3.0 \text{ mph}$$
$$- 2.5 \text{ mph}$$
$$= 64.5 \text{ mph}$$

The FFS is used as the average passenger car speed, S. The density of flow for a six-lane freeway (three lanes in each direction) is

$$D_{3 \text{ lanes}} = \frac{v_{p3 \text{ lanes}}}{S_{3 \text{ lanes}}} = \frac{1874 \text{ pcphpl}}{64.5 \dfrac{\text{mi}}{\text{hr}}}$$
$$= 29.1 \text{ pcpmpl}$$

The calculated density (29.1 pcpmpl) is greater than 26 pcpmpl (the maximum density for LOS C from the HCM LOS criteria exhibit). The expected LOS is D if only three lanes are provided in each direction. Therefore, revise the proposed design and provide four lanes in each direction and check the density again.

The change in number of lanes will also change the 15 min passenger car equivalent flow rate ($v_p = 1874$ pcphpl) because it was based on a six-lane freeway.

$$v_{p,4 \text{ lanes}} = \tfrac{3}{4} v_{p,3 \text{ lanes}}$$
$$= \left(\frac{3}{4}\right)(1874 \text{ pcphpl})$$
$$= 1406 \text{ pcphpl}$$

The FFS, assuming four lanes in each direction, can be determined from the appropriate HCM equation.

$$BFFS = 70 \text{ mph} \quad [\text{given}]$$
$$f_{LW} = 0.0 \text{ mph} \quad [\text{HCM, assuming ideal 12 ft lanes}]$$
$$f_{LC} = 0.0 \text{ mph} \quad \left[\begin{array}{c}\text{HCM, assuming ideal lateral} \\ \text{clearance} \geq 6 \text{ ft}\end{array}\right]$$
$$f_N = 1.5 \text{ mph} \quad \left[\begin{array}{c}\text{HCM, assuming four lanes} \\ \text{in each direction}\end{array}\right]$$
$$f_{ID} = 2.5 \text{ mph} \quad [\text{HCM}]$$
$$FFS = 70 \text{ mph} - 0.0 \text{ mph} - 0.0 \text{ mph} - 1.5 \text{ mph}$$
$$- 2.5 \text{ mph}$$
$$= 66 \text{ mph}$$

The density of flow for an eight-lane freeway (four lanes in each direction) is given by the HCM density equation.

$$D_{4 \text{ lanes}} = \frac{v_{p,4 \text{ lanes}}}{S_{4 \text{ lanes}}} = \frac{1406 \text{ pcphpl}}{66 \dfrac{\text{mi}}{\text{hr}}}$$
$$= 21.3 \text{ pcpmpl}$$

The calculated density (21.3 pcpmpl) is less than 26 pcpmpl (the maximum density for LOS C from the HCM exhibit). Therefore, the expected freeway LOS is C if four lanes are provided in each direction.

The answer is (C).

173. Use the HCM LOS criteria exhibit. The freeway capacity per lane when $v/c = 1.0$ is 2300 pcphpl at 60 mph free-flow speed. For a growth factor of $i = 5\%$, the number of years, n, until the freeway reaches capacity can be determined as follows.

$$v_{p,\text{future}} = v_{p,\text{present}}(1 + i)^n$$
$$2300 \text{ pcphpl} = (1900 \text{ pcphpl})(1 + 0.05)^n$$
$$\log 2300 \text{ pcphpl} = \log 1900 \text{ pcphpl} + n \log(1 + 0.05)$$
$$n = 3.9 \text{ yr} \quad (4 \text{ yr})$$

The answer is (D).

174. The horizontal distance from the BVC track centerline is

$$x_{\text{rail}} = \frac{L}{2} + (\text{sta } 28 + 50) - (\text{sta } 26 + 00)$$
$$= \frac{16.48 \text{ sta}}{2} + (\text{sta } 28 + 50) - (\text{sta } 26 + 00)$$
$$= 10.74 \text{ sta}$$

$$\text{elev}_{\text{BVC}} = \text{elev}_V - G_1 \frac{L}{2}$$
$$= 231.00 \text{ ft} - \left(3 \ \frac{\text{ft}}{\text{sta}}\right)\left(\frac{16.48 \text{ sta}}{2}\right)$$
$$= 206.28 \text{ ft}$$

The curve elevation at track centerline is

$$\text{elev}_{28+50} = \left(\frac{G_2 - G_1}{2L}\right)x_{\text{rail}}^2 + G_1 x_{\text{rail}} + \text{elev}_{\text{BVC}}$$
$$= \left(\frac{-2 \ \dfrac{\text{ft}}{\text{sta}} - \left(+3 \ \dfrac{\text{ft}}{\text{sta}}\right)}{(2)(16.48 \text{ sta})}\right)(10.74 \text{ sta})^2$$
$$+ \left(3 \ \frac{\text{ft}}{\text{sta}}\right)(10.74 \text{ sta}) + 206.28 \text{ ft}$$
$$= 221.00 \text{ ft}$$

The distance D is the difference between the curve elevation at the track center line and the elevation of the railbed at that same point.

$$D = 221.00 \text{ ft} - 195.00 \text{ ft}$$
$$= 26 \text{ ft}$$

The answer is (C).

175. There are different ways to determine the minimum length of a vertical curve. The quickest is to use the AASHTO Green Book exhibit on design controls for crest vertical curves with open road conditions.

Calculate the algebraic difference, A, in grades.

$$A = |G_2 - G_1|$$
$$= |-6\% - 2\%|$$
$$= 8\%$$

Enter a horizontal line from the algebraic difference in grades into the Green Book exhibit to intersect the 55 mph curve. Drop a vertical line from the point of intersection, and read the minimum length of curve, L, which in this case is about 900 ft.

The answer is (D).

176. The elevation of the EVC is

$$\text{elev}_{\text{EVC}} = \text{elev}_V + G_2 \frac{L}{2}$$
$$= 231.00 \text{ ft} + \left(-2 \ \frac{\text{ft}}{\text{sta}}\right)\left(\frac{16.48 \text{ sta}}{2}\right)$$
$$= 214.52 \text{ ft} \quad (210 \text{ ft})$$

The answer is (C).

177. The horizontal distance from the BVC is

$$x_{\text{high}} = \frac{LG_1}{|G_2 - G_1|}$$
$$= \frac{(22.00 \text{ sta})\left(3 \ \dfrac{\text{ft}}{\text{sta}}\right)}{\left|-5 \ \dfrac{\text{ft}}{\text{sta}} - 3 \ \dfrac{\text{ft}}{\text{sta}}\right|}$$
$$= 8.25 \text{ sta}$$

The station of the BVC is

$$\text{sta}_{\text{BVC}} = \text{sta}_V - \frac{L}{2}$$
$$= (\text{sta } 91 + 70) - \frac{22.00 \text{ sta}}{2}$$
$$= \text{sta } 80 + 70$$

The station of the highest point on the curve (the turning point) is

$$\text{sta}_{\text{high}} = \text{sta}_{\text{BVC}} + x_{\text{high}}$$
$$= (\text{sta } 80 + 70) + 8.25 \text{ sta}$$
$$= \text{sta } 88 + 95$$

The answer is (B).

178. The distance from the point of intersection of the two tangents (the vertex) to the midpoint of the curve is the external distance, E.

$$E = R\left(\sec \frac{I}{2} - 1\right)$$
$$= (1100 \text{ ft})\left(\sec \frac{45°}{2} - 1\right)$$
$$= 90.63 \text{ ft} \quad (91 \text{ ft})$$

The answer is (C).

179. The horizontal distance from the BVC is

$$x_{\text{low}} = \frac{-LG_1}{|G_2 - G_1|} = \frac{-(9 \text{ sta})\left(-4 \ \dfrac{\text{ft}}{\text{sta}}\right)}{\left|1 \ \dfrac{\text{ft}}{\text{sta}} - \left(-4 \ \dfrac{\text{ft}}{\text{sta}}\right)\right|}$$
$$= 7.20 \text{ sta}$$

The elevation of the BVC is

$$\text{elev}_{\text{BVC}} = \text{elev}_V - G_1 \frac{L}{2}$$

$$= 2231.31 \text{ ft} - \left(-4 \frac{\text{ft}}{\text{sta}}\right)\left(\frac{9.00 \text{ sta}}{2}\right)$$

$$= 2249.31 \text{ ft}$$

The elevation of the lowest point on the curve (the turning point) is

$$\text{elev}_{\text{low}} = \left(\frac{G_2 - G_1}{2L}\right) x_{\text{low}}^2 + G_1 x_{\text{low}} + \text{elev}_{\text{BVC}}$$

$$= \left(\frac{1 \frac{\text{ft}}{\text{sta}} - \left(-4 \frac{\text{ft}}{\text{sta}}\right)}{(2)(9.00 \text{ sta})}\right) (7.20 \text{ sta})^2$$

$$+ \left(-4 \frac{\text{ft}}{\text{sta}}\right)(7.20 \text{ sta}) + 2249.31 \text{ ft}$$

$$= 2234.91 \text{ ft} \quad (2235 \text{ ft})$$

The answer is (D).

180. Use the procedure for analysis of two-lane highways with general terrain since no specific grade was given. The passenger car equivalent flow rate for the peak 15 min period is given by the relevant HCM equation.

$$v_p = \frac{V}{(\text{PHF}) f_G f_{\text{HV}}}$$

The adjustment factor for grade is

$$f_G = 0.99 \quad \text{[HCM]}$$

The passenger car equivalent for heavy trucks is

$$E_T = 1.9 \quad \text{[HCM]}$$

The passenger car equivalent for RVs is

$$E_R = 1.1 \quad \text{[HCM]}$$

The adjustment factor for the presence of heavy vehicles, f_{HV}, is determined using the appropriate HCM equation.

$$f_{\text{HV}} = \frac{1}{1 + P_T(E_T - 1) + P_R(E_R - 1)}$$

$$= \frac{1}{1 + (0.08)(1.9 - 1) + (0.02)(1.1 - 1)}$$

$$= 0.931$$

$$v_p = \frac{V}{(\text{PHF}) f_G f_{\text{HV}}}$$

$$= \frac{1100 \frac{\text{veh}}{\text{hr}}}{(0.92)(0.99)\left(0.931 \frac{\text{veh}}{\text{pc}}\right)}$$

$$= 1297 \text{ pcph}$$

The highest directional flow rate for the peak 15-min period in pcph is

$$V_{p,\text{max}} = 0.6 v_p$$

$$= (0.6)(1297 \text{ pcph})$$

$$= 778 \text{ pcph}$$

The answer is (C).

181. Estimate the expected traffic volume that will use the new freeway by plotting the point of intersection of the demand and supply curves. The two curves can be combined as shown. At the point of intersection, the volume is most nearly 1750 vph.

The answer is (D).

Figure for Solution 181

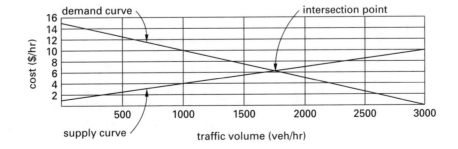

182. Total overhaul distance (OHD) is the difference between the average total haul distance (THD) and free haul distance (FHD).

$$\text{OHD} = \text{THD} - \text{FHD}$$
$$= 840 \text{ ft} - 500 \text{ ft}$$
$$= 340 \text{ ft}$$

Assuming 100 ft stations, the OHD in stations is

$$\text{OHD}_s = (340 \text{ ft})\left(\frac{1 \text{ sta}}{100 \text{ ft}}\right)$$
$$= 3.40 \text{ sta}$$

The overhaul in cubic yards (OH$_v$) is 730 yd^3 from the mass diagram for 500 ft free haul. The overhaul in cubic yard-stations (OH) is

$$\text{OH} = (\text{OH}_v)(\text{OHD}_s)$$
$$= (730 \text{ yd}^3)(3.40 \text{ sta})$$
$$= 2482 \text{ yd}^3\text{-sta}$$

The overhaul cost (OHC) depends on the overhaul unit cost (UC) and the amount of overhaul.

$$\text{OHC} = (\text{UC})(\text{OH})$$
$$= \left(9.75 \ \frac{\$}{\text{yd}^3\text{-sta}}\right)(2482 \text{ yd}^3\text{-sta})$$
$$= \$24{,}200 \quad (\$24{,}000)$$

The answer is (C).

183. The service rate per hour, Q, depends on the service time per vehicle (SPV).

$$Q = (\text{SPV})\left(\frac{60 \text{ min}}{1 \text{ hr}}\right)$$
$$= \left(\frac{1 \text{ veh}}{1.5 \text{ min}}\right)\left(\frac{60 \text{ min}}{1 \text{ hr}}\right)$$
$$= 40 \text{ vph}$$

The service rate is greater than the arrival rate, q, of 30 vph. Therefore, the queue is undersaturated. The expected number of vehicles waiting in the queue is

$$L_q = \frac{q^2}{Q(Q-q)}$$
$$= \frac{\left(30 \ \dfrac{\text{veh}}{\text{hr}}\right)^2}{\left(40 \ \dfrac{\text{veh}}{\text{hr}}\right)\left(40 \ \dfrac{\text{veh}}{\text{hr}} - 30 \ \dfrac{\text{veh}}{\text{hr}}\right)}$$
$$= 2.25$$

The answer is (B).

184. Using the gravity model and the productions, P, and attractions, A, of the related zones, the trips produced in zone 1 and attracted to zone 3 are

$$T_{1,3} = P_1\left(\frac{A_3 F_{1,3} K_{1,3}}{A_2 F_{1,2} K_{1,2} + A_3 F_{1,3} K_{1,3} + A_4 F_{1,4} K_{1,4}}\right)$$

$$= (1600 \text{ trips})\left(\frac{(1100 \text{ trips})(8)(1)}{\begin{array}{l}(840 \text{ trips})(10)(1) \\ + (1100 \text{ trips})(8)(1) \\ + (650 \text{ trips})(24)(1)\end{array}}\right)$$

$$= 429.2 \text{ trips} \quad (430 \text{ trips})$$

The factor K is assumed to be 1 because the problem stated that the socioeconomic conditions are the same for all zones.

The answer is (B).

185. Based on the node-to-node travel times from the illustration in the problem, the minimum travel times from node 1 to all the other nodes are shown in the following illustration. The links with the longer total travel times from node 1 are eliminated based on an all-or-nothing approach.

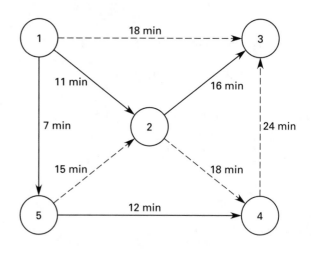

- - - longer total travel times

(not to scale)

Link 1-5 will carry the volume from node 1 to both nodes 5 and 4. The total volume, V, on link 1-5 is 245 veh (115 veh + 130 veh) as shown in the next illustration.

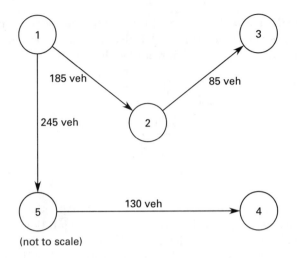

(not to scale)

The vehicle minutes of travel, VM, is the total volume, V, multiplied by the travel time, T.

$$\text{VM} = VT = (245 \text{ veh})(7 \text{ min})$$
$$= 1715 \text{ veh-min} \quad (1700 \text{ veh-min})$$

The answer is (C).

186. The QRS method is based on incorporating service elements in the estimation. Impedance of the auto mode between zones 1 and 2, $I_{1,2a}$, is a function of the in-vehicle time, VT_a, in minutes; the excess time, ET_a, in minutes; the trip cost, TC_a; and the income per min, IN_a.

$$I_{1,2a} = \text{VT}_a + 2.5\text{ET}_a + 3\frac{\text{TC}_a}{\text{IN}_a}$$

$$= \left(\frac{12 \text{ mi}}{55\frac{\text{mi}}{\text{hr}}}\right)\left(60\frac{\text{min}}{\text{hr}}\right) + (2.5)(6 \text{ min})$$

$$+ (3)\left(\frac{\left(0.35\frac{\$}{\text{mi}}\right)(12 \text{ mi})}{\frac{\$30,000}{120,000 \text{ min}}}\right)$$

$$= 78.49 \text{ min}$$

Similarly, impedance of the transit mode between zones 1 and 2 is

$$I_{1,2t} = \text{VT}_t + 2.5\text{ET}_t + 3\frac{\text{TC}_t}{\text{IN}_t}$$

$$= \left(\frac{10 \text{ mi}}{45\frac{\text{mi}}{\text{hr}}}\right)\left(60\frac{\text{min}}{\text{hr}}\right) + (2.5)(11 \text{ min})$$

$$+ (3)\left(\frac{\left(0.20\frac{\$}{\text{mi}}\right)(10 \text{ mi})}{\frac{\$30,000}{120,000 \text{ min}}}\right)$$

$$= 64.83 \text{ min}$$

The percentage of trips between zones 1 and 2 that are expected to use auto, MS_a, is specified by the QRS model.

$$\text{MS}_a = \frac{I_{1,2a}^b}{I_{1,2t}^b + I_{1,2a}^b} \times 100\%$$

$$= \frac{(78.49 \text{ min})^{1.5}}{(64.83 \text{ min})^{1.5} + (78.49 \text{ min})^{1.5}} \times 100\%$$

$$= 57\%$$

The answer is (B).

187. The expected number of fatal accidents after the development, FA_a, is the number of fatal accidents before the development, FA_b, reduced by 25%.

$$\text{FA}_a = (1 - \text{reduction})\text{FA}_b$$
$$= (1 - 0.25)(140 \text{ total accidents})$$
$$\times \left(\frac{5 \text{ fatal accidents}}{100 \text{ total accidents}}\right)$$
$$= 5.25 \text{ fatal accidents}$$

The vehicle-miles of travel (VMT) during the 3 yr period, T, after the development and improvements on this 15 mile section, L, is

$$\text{VMT} = (\text{ADT})TL$$
$$= \left(18,000\frac{\text{veh}}{\text{day}}\right)(3 \text{ yr})\left(365\frac{\text{days}}{\text{year}}\right)(15 \text{ mi})$$
$$= 295.65 \times 10^6 \text{ veh-mi}$$

The fatal accident rate, RPVM, per 100 million veh-mi of travel (HMVM) is

$$\text{RPVM} = \frac{\text{FA}_a(100 \times 10^6)}{\text{VMT}}$$
$$= \frac{(5.25 \text{ fatal accidents})(100 \times 10^6)}{295.65 \times 10^6 \text{ veh-mi}}$$
$$= 1.78 \text{ fatal accidents/HMVM}$$
$$(1.8 \text{ fatal accidents/HMVM})$$

The answer is (B).

188. The total number of vehicles entering the intersections is

$$\text{VE} = \sum \text{entering traffic}$$
$$= 1250\frac{\text{veh}}{\text{day}} + 2350\frac{\text{veh}}{\text{day}} + 730\frac{\text{veh}}{\text{day}} + 1920\frac{\text{veh}}{\text{day}}$$
$$= 6250 \text{ veh/day}$$

From the problem statement, 13 injury-causing accidents, IA, occurred in 1 yr. The rate of IA per 10 million entering vehicles is

$$\text{RPEV} = \frac{\text{IA}(10 \times 10^6)}{\text{VE}\left(365 \, \dfrac{\text{day}}{\text{yr}}\right)}$$

$$= \frac{\left(13 \, \dfrac{\text{injury accidents}}{\text{yr}}\right)(10 \times 10^6)}{\left(6250 \, \dfrac{\text{veh}}{\text{day}}\right)\left(365 \, \dfrac{\text{day}}{\text{year}}\right)}$$

$$= 57 \text{ accidents}/10^6 \text{ veh}$$

The answer is (B).

189. Based on the percent of commuters, P_1, the number of commuting vehicles expected to use the parking garage is

$$V_1 = P_1V = (0.75)(400 \text{ veh})$$
$$= 300 \text{ veh}$$

Based on the percent of shoppers, P_2, the number of shopping vehicles expected to use the parking garage is

$$V_2 = P_2V = (0.25)(400 \text{ veh})$$
$$= 100 \text{ veh}$$

The total demand for the garage, D, is based on the number of vehicles and the time they spend parked. t_1 represents the average parking duratuion of the commuters, and t_2 is the average parking duration of the shoppers.

$$D = V_1t_1 + V_2t_2$$
$$= (300 \text{ veh})(8 \text{ hr}) + (100 \text{ veh})(3 \text{ hr})$$
$$= 2700 \text{ veh-hr}$$

Because each vehicle is expected to occupy one space, the demand can be expressed as 2700 spaces-hr. The total time the parking garage is available for parking, T, is from 6 a.m. to 5 p.m. (11 hr). f is the parking efficiency. N is the number of spaces required to meet the demand. The demand is equal to the product of the efficiency, the available time, and the number of spaces required to meet demand. The required supply, S, should be equal to or greater than the demand.

$$S \geq D = fTN$$
$$2700 \text{ spaces-hr} = (0.85)(11 \text{ hr})N$$
$$N = 289 \text{ spaces} \quad (290 \text{ spaces})$$

The answer is (C).

190. The width of available space for parking on each side of the road, W_P, is calculated from the width of the road, W, and the width of each lane, W_L.

$$W_P = \frac{W - 2W_L}{2}$$
$$= \frac{40 \text{ ft} - (2)(11 \text{ ft})}{2}$$
$$= 9 \text{ ft}$$

This parking width (9 ft) can only accommodate parallel parking, which is the parking configuration with the least traffic interference. The length of a parking space in a parallel parking configuration is based on a standard of 22 ft.

The total number of parallel parking spaces on both sides of the road is

$$N = \frac{(\text{no. of sides})(\text{road length})}{\text{stall length}}$$

$$= \frac{(2)(1.2 \text{ mi})\left(\dfrac{5280 \text{ ft}}{1 \text{ mi}}\right)}{22 \, \dfrac{\text{ft}}{\text{space}}}$$

$$= 576 \text{ spaces} \quad (580 \text{ spaces})$$

The answer is (B).

191. The weaving section is of type A configuration according to the *Highway Capacity Manual* (HCM) because both weaving movements must make one lane change.

The larger of the nonweaving flows is

$$v_1 = 2500 \, \frac{\text{passenger cars}}{\text{hr}} + 2000 \, \frac{\text{passenger cars}}{\text{hr}}$$
$$- 200 \, \frac{\text{passenger cars}}{\text{hr}}$$
$$= 4300 \text{ passenger cars/hr}$$

The smaller of the nonweaving flows is

$$v_2 = 450 \, \frac{\text{passenger cars}}{\text{hr}} - 300 \, \frac{\text{passenger cars}}{\text{hr}}$$
$$= 150 \text{ passenger cars/hr}$$

The weaving flows are shown in the weaving diagram.

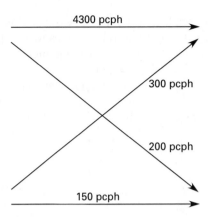

4300 pcph

300 pcph

200 pcph

150 pcph

The critical ratios can be calculated based on the weaving diagram and the definitions of critical ratios given in HCM exhibits.

Total weaving flow rate in the weaving area is

$$v_w = 200 \; \frac{\text{passenger cars}}{\text{hr}} + 300 \; \frac{\text{passenger cars}}{\text{hr}}$$
$$= 500 \; \text{passenger cars/hr}$$

Total flow in the weaving area is

$$v = 4300 \; \frac{\text{passenger cars}}{\text{hr}} + 500 \; \frac{\text{passenger cars}}{\text{hr}}$$
$$+ 150 \; \frac{\text{passenger cars}}{\text{hr}}$$
$$= 4950 \; \text{passenger cars/hr}$$

The volume ratio is

$$\text{VR} = \frac{v_w}{v}$$
$$= \frac{500 \; \dfrac{\text{passenger cars}}{\text{hr}}}{4950 \; \dfrac{\text{passenger cars}}{\text{hr}}}$$
$$= 0.101$$

The weaving section has a total of three lanes, N, and a length of 1500 ft, L. The weaving intensity factor, W_i, is given by an HCM equation. The constants a, b, c, and d are obtained from HCM.

$$W_i = \frac{a(1 + \text{VR})^b \left(\dfrac{v}{N}\right)^c}{L^d}$$
$$= \frac{(0.15)(1 + 0.101)^{2.2} \left(\dfrac{4950 \; \dfrac{\text{passenger cars}}{\text{hr-lane}}}{3 \; \text{lanes}}\right)^{0.97}}{(1500 \; \text{ft})^{0.80}}$$
$$= 0.705$$

The weaving speed for unconstrained operations is given by the relevant HCM equation.

$$S_w = 15 \; \frac{\text{mi}}{\text{hr}} + \frac{S_{\text{FF}} - 10 \; \dfrac{\text{mi}}{\text{hr}}}{1 + W_i}$$
$$= 15 \; \frac{\text{mi}}{\text{hr}} + \frac{60 \; \dfrac{\text{mi}}{\text{hr}} - 10 \; \dfrac{\text{mi}}{\text{hr}}}{1 + 0.705}$$
$$= 44.3 \; \text{mph} \quad (44 \; \text{mph})$$

The answer is (B).

192. The total lost time for the whole intersection, L, is the sum of lost times for all the phases.

$$L = \sum t_l$$
$$= 3 \; \text{sec} + 3 \; \text{sec} + 3 \; \text{sec} + 3 \; \text{sec}$$
$$= 12 \; \text{sec}$$

Using the appropriate HCM equation, the cycle length of the intersection traffic signal, C, can be determined given the desired critical ratio of flow to capacity for the overall intersection, X_c.

$$X_c = \sum \left(\frac{v}{s}\right)_{ci} \left(\frac{C}{C - L}\right)$$
$$0.90 = (0.35 + 0.10 + 0.23 + 0.08) \left(\frac{C}{C - 12 \; \text{sec}}\right)$$
$$C = 77.1 \; \text{sec} \quad (78 \; \text{sec})$$

The answer is (B).

193. The problem statement provides information about the intersection: the maximum allowable speed, v; the width of intersection, W; the average length of vehicle, L; the perception-reaction time, t_{PR}; and the deceleration rate, a.

The minimum yellow interval at this intersection is

$$t_{\text{yellow,min}} = t_{\text{PR}} + \frac{W + L}{u} + \frac{\text{v}}{2a}$$
$$= 2 \; \text{sec} + \frac{48 \; \text{ft} + 20 \; \text{ft}}{\left(45 \; \dfrac{\text{mi}}{\text{hr}}\right)\left(\dfrac{5280 \; \text{ft}}{1 \; \text{mi}}\right)\left(\dfrac{1 \; \text{hr}}{3600 \; \text{sec}}\right)}$$
$$+ \frac{\left(45 \; \dfrac{\text{mi}}{\text{hr}}\right)\left(\dfrac{5280 \; \text{ft}}{1 \; \text{mi}}\right)\left(\dfrac{1 \; \text{hr}}{3600 \; \text{sec}}\right)}{(2)\left(11.2 \; \dfrac{\text{ft}}{\text{sec}^2}\right)}$$
$$= 5.98 \; \text{sec} \quad (6.0 \; \text{sec})$$

The answer is (D).

194. Compound curves consist of two or more curves turning in the same direction while reverse curves turn in opposite directions. Therefore, the statement in option (C) is not true. All the other statements are true.

The answer is (C).

195. Rapid-curing asphalt is formed by cutting asphalt cement with a petroleum distillate, such as gasoline, that will easily evaporate.

The answer is (C).

196. A soil classified as A-7-6 (20) is usually rated "poor" as a subgrade and is considered unsuitable as a subbase material. Therefore, the statement in option (D) is not true. All other statements are true.

The answer is (D).

197. From the illustrations, for maximum unit weight, the asphalt content AC_{UW} is 5.4%. For maximum marshal stability, the asphalt content AC_{ST} is 4.9%. For 4% total air voids (VTM), the asphalt content AC_{AI} is 4.5%. The optimum asphalt content AC_{OP} is

$$AC_{OP} = \frac{AC_{UW} + AC_{ST} + AC_{AI}}{3}$$
$$= \frac{5.4\% + 4.9\% + 4.5\%}{3}$$
$$= 4.93\% \quad (5\%)$$

The answer is (C).

198. The problem statement provides information about the paving mixture: the bulk specific gravity of the combined aggregate in the mixture, G_{sb} the mass of the core in air, A; and the mass of the submerged core, C.

The bulk specific gravity of the mix is

$$G_{mb} = \frac{A}{A - C}$$
$$= \frac{1238.5 \text{ g}}{1238.5 \text{ g} - 698.3 \text{ g}}$$
$$= 2.29$$

The aggregate percent by weight of total paving mixture, P_s, is $100\% - 6\% = 94\%$. The percent VMA is

$$VMA = 100\% - \frac{G_{mb}P_s}{G_{sb}}$$
$$= 100\% - \frac{(2.29)(94\%)}{2.64}$$
$$= 18.46\% \quad (18\%)$$

The answer is (C).

199. The problem statement gives the following information about the project location and the pavement design: latitude, lat; the 7 d average high air temperature, T_{high}; the 1 d average low air temperature, T_{low}; the standard deviation of the high temperature, σ_{high}; the standard deviation of the low temperature, σ_{low}; the pavement surface depth, H; and the reliability.

The 7 d average high air temperature for 98% ($z = 2.055$) reliability is

$$T_{high98} = T_{high} + 2.055\sigma_{high} = 31°C + (2)(1.5°C)$$
$$= 34°C$$

The design pavement temperature at a depth of 20 mm is

$$T_{20 \text{ mm}} = \begin{pmatrix} T_{high98} - 0.00618(\text{lat})^2 \\ + 0.2289(\text{lat}) + 42.2 \end{pmatrix}(0.9545) - 17.78$$
$$= \begin{pmatrix} 34°C - (0.00618)(40.9°)^2 \\ + (0.2289)(40.9°) \\ + 42.2 \end{pmatrix}(0.9545) - 17.78$$
$$= 54°C$$

The low pavement design temperature is

$$T_{pave} = 1.56 + 0.72T_{low} - 0.004(\text{lat})^2$$
$$+ 6.26\log_{10}(H + 25)$$
$$- z\sqrt{4.4 + 0.52\sigma_{low}^2}$$
$$= 1.56 + (0.72)(-25°C) - (0.004)(40.9°)^2$$
$$+ 6.26\log_{10}(190 \text{ mm} + 25)$$
$$- 2.055\sqrt{4.4 + (0.52)(2.7°)^2}$$
$$= -14.4°C \quad (-14°C)$$

From a table that contains performance-graded asphalt binder specifications, use the computed values of $T_{20 \text{ mm}}$ and T_{pave} to select an appropriate performance grade asphalt binder for this project. The asphalt binder PG 58-16 is the most appropriate for 98% reliability.

The answer is (C).

200. The problem statement gives information about the constructed pavement, including its thickness, D; the layer coefficient, a; and the drainage coefficient, m. The materials used in the different layers—AC surface course, untreated granular base, and untreated gravel subbase—are represented by the subscripts 1, 2, and 3, respectively.

material	thickness, D (in)	layer coefficient, a	drainage coefficient, m
AC surface course, D_1	6	0.400	–
untreated granular base, D_2	8	0.115	0.50
untreated gravel subbase, D_3	10	0.090	0.50

The SN of the constructed pavement is

$$\begin{aligned} \text{SN} &= a_1 D_1 + a_2 D_2 m_2 + a_3 D_3 m_3 \\ &= (0.400)(6 \text{ in}) + (0.115)(8 \text{ in})(0.50) \\ &\quad + (0.09)(10 \text{ in})(0.50) \\ &= 3.31 \quad (3.3) \end{aligned}$$

The answer is (B).

Solutions
Water Resources

201. The ratio of the design head to the test head is

$$\frac{H}{H_o} = \frac{16 \text{ ft}}{10 \text{ ft}} = 1.6$$

From the test results, the coefficient of discharge is

$$\frac{C}{C_o} = 1.09$$
$$C = 1.09 C_o$$
$$= (1.09)\left(2.20 \, \frac{\text{ft}^{1/2}}{\text{sec}}\right)$$
$$= 2.40 \text{ ft}^{1/2}/\text{sec}$$

The discharge for a broad-crested weir is

$$Q = C b H^{3/2}$$
$$= \left(2.40 \, \frac{\text{ft}^{1/2}}{\text{sec}}\right)(16 \text{ ft})(16 \text{ ft})^{3/2}$$
$$= 2457 \text{ ft}^3/\text{sec} \quad (2500 \text{ ft}^3/\text{sec})$$

The answer is (B).

202. The statements are all true.

The answer is (B).

203. The total available energy head is equal to the velocity head at the discharge plus the total losses in the piping system. The losses include friction at the square entrance, h_e; friction in pipe section 1-2, h_{f1}; sudden contraction, h_c; friction in pipe section 2-3, h_{f2}; friction in the rotary valve, h_v; and the velocity head at the free discharge, h_d. The minor losses are expressed using a loss coefficient, K, of the velocity head. The head loss for a square edge entrance ($K_e = 0.5$) is

$$h_e = K_e h_v = K_e \frac{\text{v}_1^2}{2 g_c}$$
$$= 0.5 \frac{\text{v}_1^2}{2 g_c}$$

The head loss for pipe section 1-2 is

$$h_{f_1} = f_1 \left(\frac{L}{D}\right)\left(\frac{\text{v}_1^2}{2g}\right)$$
$$= f_1 \left(\frac{3940 \text{ ft}}{20 \text{ in}}\right)\left(\frac{\text{v}_1^2}{2g}\right)\left(12 \, \frac{\text{in}}{\text{ft}}\right)$$
$$= 2364 f_1 \frac{\text{v}_1^2}{2g}$$

The head loss for the sudden contraction is

$$K_c = \frac{1}{2}\left(1 - \left(\frac{D_s}{D_l}\right)^2\right)$$
$$= \left(\frac{1}{2}\right)\left(1 - \left(\frac{12 \text{ in}}{20 \text{ in}}\right)^2\right)$$
$$= 0.32$$
$$h_c = 0.32 \frac{\text{v}_2^2}{2g}$$

The head loss for pipe section 2-3 is

$$h_{f_2} = f_2 \left(\frac{L}{D}\right)\left(\frac{\text{v}_2^2}{2g}\right)$$
$$= f_2 \left(\frac{4920 \text{ ft}}{12 \text{ in}}\right)\left(\frac{\text{v}_2^2}{2g}\right)\left(12 \, \frac{\text{in}}{\text{ft}}\right)$$
$$= 4920 f_2 \frac{\text{v}_2^2}{2g}$$

The head loss for a fully open rotary valve is

$$h_v = \frac{10 \text{v}_2^2}{2g}$$

The head loss for the free discharge is

$$h_d = \frac{1.0 \text{v}_2^2}{2g}$$

From the continuity equation,

$$\text{v}_1 = \frac{A_2 \text{v}_2}{A} = \frac{D_2^2 \text{v}_2}{D_1^2}$$
$$= \frac{(12 \text{ in})^2 \text{v}_2}{(20 \text{ in})^2}$$
$$= 0.36 \text{v}_2$$

To evaluate f_1 and f_2, the Reynolds numbers for points 1 and 2 are needed.

$$Re_1 = \frac{D_1 v_1}{\nu} = \frac{(20 \text{ in})\left(\dfrac{1 \text{ ft}}{12 \text{ in}}\right) v_1}{1.410 \times 10^{-5} \dfrac{\text{ft}^2}{\text{sec}}}$$

$$= 1.18 \times 10^5 v_1$$

$$Re_2 = \frac{D_2 v_2}{\nu} = \frac{(12 \text{ in})\left(\dfrac{1 \text{ ft}}{12 \text{ in}}\right) v_2}{1.410 \times 10^{-5} \dfrac{\text{ft}^2}{\text{sec}}}$$

$$= 7.09 \times 10^4 v_2$$

The relative roughness for each of the two pipes is

$$\frac{\epsilon_1}{D_1} = \left(\frac{0.004 \text{ ft}}{20 \text{ in}}\right)\left(12 \frac{\text{in}}{\text{ft}}\right)$$

$$= 0.0024$$

$$\frac{\epsilon_2}{D_2} = \left(\frac{0.004 \text{ ft}}{12 \text{ in}}\right)\left(12 \frac{\text{in}}{\text{ft}}\right)$$

$$= 0.004$$

Solve by iteration. For trial 1, assume f_1 is 0.022 and f_2 is 0.028. The head loss relationships are

$$\Delta E_k = \Delta E_p$$

$$h_e + h_{f1} + h_c + h_{f2} + h_v + h_d = 295 \text{ ft}$$

$$(0.5 + 2364 f_1)\left(\frac{0.36 v_2^2}{(2)\left(32.2 \dfrac{\text{ft}}{\text{sec}^2}\right)}\right)$$

$$+ (0.32 + 4920 f_2 + 10 + 1)$$

$$\times \left(\frac{v_2^2}{(2)\left(32.2 \dfrac{\text{ft}}{\text{sec}^2}\right)}\right) = 295 \text{ ft}$$

Solve for v_2 using a calculator solver, with $f_1 = 0.022$ and $f_2 = 0.028$.

$$v_2 = 10.63 \text{ ft/sec}$$

$$v_1 = 0.36 v_2$$

$$= (0.36)\left(10.63 \frac{\text{ft}}{\text{sec}}\right)$$

$$= 3.83 \text{ ft/sec}$$

The Reynolds numbers are

$$Re_1 = \frac{(20 \text{ in})\left(\dfrac{1 \text{ ft}}{12 \text{ in}}\right)\left(3.83 \dfrac{\text{ft}}{\text{sec}}\right)}{1.410 \times 10^{-5} \dfrac{\text{ft}^2}{\text{sec}}}$$

$$= 4.53 \times 10^5$$

$$Re_2 = \frac{(12 \text{ in})\left(\dfrac{1 \text{ ft}}{12 \text{ in}}\right)\left(10.63 \dfrac{\text{ft}}{\text{sec}}\right)}{1.410 \times 10^{-5} \dfrac{\text{ft}^2}{\text{sec}}}$$

$$= 7.54 \times 10^5$$

From a Moody diagram, the friction factors are $f_1 = 0.025$ and $f_2 = 0.028$. These values are close enough to the assumed friction factors used to find the pipe velocities. The velocities in the pipe are

$$v_1 = 4 \text{ ft/sec}$$

$$v_2 = 11 \text{ ft/sec}$$

The discharge is

$$Q = A_2 v_2 = \frac{D_2^2}{4} v_2$$

$$= \frac{\pi \left((12 \text{ in})\left(\dfrac{1 \text{ ft}}{12 \text{ in}}\right)\right)^2 \left(11 \dfrac{\text{ft}}{\text{sec}}\right)}{4}$$

$$= 8.6 \text{ ft}^3/\text{sec}$$

The answer is (B).

204. The difference between the water elevation in a reservoir and the water level in a piezometer at the junction is the head loss for the flow in the associated pipe. The sum of the flows into the junction must equal the sum of the flows leaving the junction.

$$\sum Q = 0$$

Also, the piezometric head at the junction is the same for all pipes that meet at the junction. This requires an iterative solution solved by assuming a piezometric head at the junction, calculating the head losses in each pipe, then solving for the flow in each pipe. The direction of flow is based on the assumed head at the junction and the flows summed accordingly. The error in the trial indicates the direction the assumed head at the junction should be set for the next trial. The friction loss in each pipe is

$$h_f = f\left(\frac{L}{D}\right)\left(\frac{v^2}{2g}\right)$$

$$= (0.04)\left(\frac{11{,}480 \text{ ft}}{1.33 \text{ ft}}\right)\left(\frac{v_1^2}{(2)\left(32.2 \dfrac{\text{ft}}{\text{sec}^2}\right)}\right)$$

$$h_{f_1} = 5.361 v_1^2$$

$$h_{f_2} = (0.04)\left(\frac{2950 \text{ ft}}{1 \text{ ft}}\right)\left(\frac{v_2^2}{(2)\left(32.2 \dfrac{\text{ft}}{\text{sec}^2}\right)}\right)$$

$$= 1.832 v_2^2$$

$$h_{f_3} = (0.04) \left(\frac{5900 \text{ ft}}{1.67 \text{ ft}} \right) \left(\frac{v_3^2}{(2) \left(32.2 \dfrac{\text{ft}}{\text{sec}^2} \right)} \right)$$

$$= 2.194 v_3^2$$

Assume the piezometric head at the junction is at an elevation of 360 ft. Other initial elevation assumptions could be made. This means Q_2 flows into the junction, and Q_1 and Q_3 flow out of the junction. For pipe 1, the head is

$$h_{f_1} = 360.0 \text{ ft} - 328.0 \text{ ft} = 32 \text{ ft}$$

The velocity in pipe 1 is

$$v_1 = \sqrt{\frac{h_{f_1}}{5.361}} = \sqrt{\frac{32 \text{ ft}}{5.361}}$$
$$= 2.443 \text{ ft/sec}$$

The flow in pipe 1 is

$$Q_1 = A_1 v_1$$
$$= (1.396 \text{ ft}^2) \left(2.443 \frac{\text{ft}}{\text{sec}} \right)$$
$$= 3.410 \text{ ft}^3/\text{sec}$$

For pipe 2, the head is

$$h_{f_2} = 426.5 \text{ ft} - 360.0 \text{ ft}$$
$$= 66.5 \text{ ft}$$

The velocity in pipe 2 is

$$v_2 = \sqrt{\frac{h_{f_2}}{1.832}} = \sqrt{\frac{66.5 \text{ ft}}{1.832}}$$
$$= 6.025 \text{ ft/sec}$$

The flow in pipe 2 is

$$Q_2 = A_2 v_2$$
$$= (0.785 \text{ ft}^2) \left(6.025 \frac{\text{ft}}{\text{sec}} \right)$$
$$= 4.730 \text{ ft}^3/\text{sec}$$

For pipe 3, the head is

$$h_{f_3} = 360.0 \text{ ft} - 295.3 \text{ ft}$$
$$= 64.7 \text{ ft}$$

The velocity in pipe 3 is

$$v_3 = \sqrt{\frac{h_{f_3}}{2.194}} = \sqrt{\frac{64.7 \text{ ft}}{2.194}}$$
$$= 5.430 \text{ ft/sec}$$

The flow in pipe 3 is

$$Q_3 = A_3 v_3$$
$$= (2.182 \text{ ft}^2) \left(5.430 \frac{\text{ft}}{\text{sec}} \right)$$
$$= 11.849 \text{ ft}^3/\text{sec}$$

The sum of flows around the junction is

$$Q_2 - Q_1 - Q_3 = 0$$
$$4.730 \frac{\text{ft}^3}{\text{sec}} - 3.410 \frac{\text{ft}^3}{\text{sec}} - 11.849 \frac{\text{ft}^3}{\text{sec}} = -10.529 \text{ ft}^3/\text{sec}$$

Since the sum of flows does not equal zero, another trial is necessary. The assumed piezometric head at the junction is too high because the flows out of the junction are too large. Assume the piezometric head at the junction is 321.0 ft. This means Q_1 and Q_2 flow into the junction, and Q_3 flows out.

$$Q_1 + Q_2 - Q_3 = 0$$

The head and velocity for pipe 1 are

$$h_{f_1} = 328.0 \text{ ft} - 321.0 \text{ ft} = 7 \text{ ft}$$
$$v_1 = \sqrt{\frac{7 \text{ ft}}{5.361}}$$
$$= 1.143 \text{ ft/sec}$$

The flow in pipe 1 is

$$Q_1 = (1.396 \text{ ft}^2) \left(1.143 \frac{\text{ft}}{\text{sec}} \right)$$
$$= 1.596 \text{ ft}^3/\text{sec}$$

For pipe 2,

$$h_{f_2} = 426.5 \text{ ft} - 321.0 \text{ ft} = 105.5 \text{ ft}$$
$$v_2 = \sqrt{\frac{105.5 \text{ ft}}{1.832}}$$
$$= 7.59 \text{ ft/sec}$$
$$Q_2 = (0.785 \text{ ft}^2) \left(7.59 \frac{\text{ft}}{\text{sec}} \right)$$
$$= 5.958 \text{ ft}^3/\text{sec}$$

For pipe 3,

$$h_{f_3} = 321.0 \text{ ft} - 295.3 \text{ ft} = 25.7 \text{ ft}$$
$$v_3 = \sqrt{\frac{25.7 \text{ ft}}{2.194}}$$
$$= 3.422 \text{ ft/sec}$$
$$Q_3 = (2.182 \text{ ft}^2) \left(3.422 \frac{\text{ft}}{\text{sec}} \right)$$
$$= 7.467 \text{ ft}^3/\text{sec}$$

The sum of flows around the junction is

$$Q_1 + Q_2 - Q_3 = 0$$

$$1.596 \ \frac{\text{ft}^3}{\text{sec}} + 5.958 \ \frac{\text{ft}^3}{\text{sec}} - 7.467 \ \frac{\text{ft}^3}{\text{sec}}$$

$$= 0.087 \ \text{ft}^3/\text{sec} \quad [\text{close enough}]$$

$$Q_1 = 1.6 \ \text{ft}^3/\text{sec into the junction}$$
$$Q_2 = 6.0 \ \text{ft}^3/\text{sec into the junction}$$
$$Q_3 = 7.5 \ \text{ft}^3/\text{sec into the junction}$$

The answer is (B).

205. The slope of the stream is

$$S_o = \frac{\Delta z}{L} = \frac{1.54 \ \text{ft}}{2135.69 \ \text{ft}}$$
$$= 0.00072 \ \text{ft/ft}$$

The Manning roughness coefficient, n, for a natural channel with stones and weeds is 0.035. The maximum flow area is

$$A = wd = (4 \ \text{ft})(6 \ \text{ft})$$
$$= 24 \ \text{ft}^2$$

The wetted perimeter is

$$P = w + 2d = 4 \ \text{ft} + (2)(6 \ \text{ft})$$
$$= 16 \ \text{ft}$$

The hydraulic radius is

$$R = \frac{A}{P} = \frac{24 \ \text{ft}^2}{16 \ \text{ft}}$$
$$= 1.5 \ \text{ft}$$

The flow from the Chezy-Manning equation is

$$Q = \frac{1.49}{n} A R^{2/3} \sqrt{S_o}$$
$$= \left(\frac{1.49}{0.035} \right)(24 \ \text{ft}^2)(1.5 \ \text{ft})^{2/3} \sqrt{0.00072}$$
$$= 35.9 \ \text{ft}^3/\text{sec} \quad (36 \ \text{ft}^3/\text{sec})$$

The answer is (A).

206. Statement III is false. Detention basins in the lower part of a river basin are most effective in reducing the flood crest of a downstream-moving storm.

Statements I, II, and IV are true.

The answer is (B).

207. Determine the system head curves for the two clear well elevations. Then, plot the pump curve against the system head curves to determine the range of operating points. The pipe friction loss is given by the Hazen-Williams equation.

$$h_f = \frac{10.44 L Q_{\text{gpm}}^{1.85}}{C^{1.85} d_{\text{inches}}^{4.8655}}$$

For design, use a roughness coefficient of 140. The discharge pipe friction loss is

$$h_{f_d} = \frac{(10.44)(2000 \ \text{ft}) Q_{\text{gpm}}^{1.85}}{(140)^{1.85}(12 \ \text{in})^{4.8655}}$$
$$= 12.55 \times 10^{-6} Q_{\text{gpm}}^{1.85} \ \text{ft}$$

The suction pipe friction loss is

$$h_{f_s} = \frac{(10.44)(1500 \ \text{ft}) Q_{\text{gpm}}^{1.85}}{(140)^{1.85}(16 \ \text{in})^{4.8655}}$$
$$= 2.322 \times 10^{-6} Q_{\text{gpm}}^{1.85} \ \text{ft}$$

At the various flows, the discharge and suction friction losses are

flow rate (gpm)	pump total dynamic head (ft)	discharge friction loss (ft)	suction friction loss (ft)
500	96	1.24	0.23
1000	88	4.45	0.83
1500	76	9.43	1.74
2000	60	16.05	2.97
2500	36	24.36	4.49

The total dynamic head is

$$h_t = h_{z_d} - h_{z_s} + h_{f_d} + h_{f_s}$$

The discharge static head is

$$h_{z_d} = 160 \ \text{ft} - 90 \ \text{ft}$$
$$= 70 \ \text{ft}$$

The suction static head at low level is

$$h_{z_s} = 100 \ \text{ft} - 90 \ \text{ft} = 10 \ \text{ft}$$

The total dynamic head at 500 gpm is

$$h_t = 70 \ \text{ft} - 10 \ \text{ft} + 1.24 \ \text{ft} + 0.23 \ \text{ft}$$
$$= 61.47 \ \text{ft} \quad [\text{low level}]$$

The total dynamic heads for other flows are given in the following table.

flow rate (gpm)	pump total dynamic head (ft)	system head head at low level (ft)	system head head at high level (ft)
500	96	61.47	41.47
1000	88	65.28	45.28
1500	76	71.17	51.17
2000	60	79.02	59.02
2500	36	88.75	68.75

Plot the flow rate against the pump and system heads.

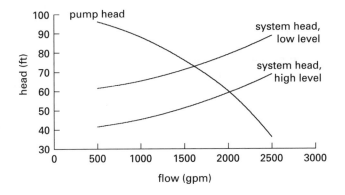

The pump operating points are 1650 gpm at 72 ft and 2000 gpm at 60 ft.

The answer is (D).

208. Statement II is false. This condition would only be true if the flows in pipes 3 and 7 were zero. Statement III is false. The pressure head at node 2 will be the same as the water surface elevation of 100 ft. Statement I is true. Statement IV is true, because the algebraic sum of the energy losses around any loop must be zero.

The answer is (C).

209. The pipe protrudes into the retention pond by two pipe diameters, which approximates a reentrant tube with a coefficient of discharge, C_d, of 0.72. The discharge from the pipe is

$$\dot{V} = C_d A_o \sqrt{2gh}$$

The flow over the spillway is

$$Q = \tfrac{2}{3} C_1 b \sqrt{2g} H^{3/2}$$

Set the flow from the pipe equal to the flow from the spillway, and solve for the head.

$$H = h - 2.5 \text{ ft}$$
$$C_d A_o \sqrt{2gh} = \tfrac{2}{3} C_1 b \sqrt{2g}(h - 2.5 \text{ ft})^{3/2}$$
$$(0.72)\left(\frac{\pi(1 \text{ ft})^2}{4}\right)\sqrt{2gh} = \left(\frac{2}{3}\right)(0.62)(4 \text{ ft})\sqrt{2g}$$
$$\times (h - 2.5 \text{ ft})^{3/2}$$
$$\frac{\sqrt{h}}{(h - 2.5 \text{ ft})^{3/2}} = 2.924$$

Solve by iteration or a calculator solver function.

$$h = 3.22 \text{ ft} \quad (3.2 \text{ ft})$$

The answer is (C).

210. The maximum daily demand for residential/commercial use is

$$Q_{\text{res}} = UPM$$

U is the per capita daily demand, P is the population, and M is the demand multiplier.

$$Q_{\text{res}} = \left(\frac{180 \text{ gal}}{\text{capita-day}}\right)(9000 \text{ capita})$$
$$\times (1.8)\left(\frac{1 \text{ day}}{1440 \text{ min}}\right)$$
$$= 2025 \text{ gpm}$$

The fire demand for essential service structures is

$$Q_{\text{fire}} = 18F\sqrt{A_{\text{ft}^2}}$$

Note that the fire flow formula is empirical and not dimensionally consistent. For zone 1, F is 0.8 for class 4. The zone 1 fire demand is

$$Q_{\text{fire}} = (18)(0.8)\sqrt{20,000 \text{ ft}^2}$$
$$= 2036 \text{ gpm}$$

Rounding to the nearest 250 gpm, the fire demand is 2000 gpm. Check that the fire flow is between 500 gpm and 6000 gpm for class 3, 4, 5, and 6 structures. It is, so use 2000 gpm. For zone 2, F is 0.6 for class 5. The zone 2 fire demand is

$$Q_{\text{fire}} = (18)(0.6)\sqrt{50,500 \text{ ft}^2}$$
$$= 2427 \text{ gpm}$$

Rounding to the nearest 250 gpm, the fire demand is 2500 gpm. Check that the fire flow is between 500 gpm and 6000 gpm for class 3, 4, 5, and 6 structures. It is, so use 2500 gpm. For zone 3, F is 0.8 for class 3. The zone 3 fire demand is

$$Q_{\text{fire}} = (18)(0.8)\sqrt{40,800 \text{ ft}^2}$$
$$= 2908 \text{ gpm}$$

Rounding to the nearest 250 gpm, the fire demand is 3000 gpm. Check that the fire flow is between 500 and 6000 gpm for class 3, 4, 5, and 6 structures. It is, so use 3000 gpm. The fire demand based on structure separations for zone 1 (25 ft separation) is 1000 gpm. The fire demand based on structure separations for zone 2 (40 ft separation) is 750 gpm. The fire demand based on structure separations for zone 3 (120 ft separation) is 500 gpm. The controlling fire demand will be the demand for essential service structures. The maximum daily demands are given in the following table.

demand type	maximum daily demand (gpm)		
	zone 1	zone 2	zone 3
residential/ commercial	2025	2025	2025
industrial	80	80	80
fire	2000	2500	3000
maximum	4105	4605	5105

The maximum demand for all uses is zone 3 with 5105 gpm. The required delivery capability that best meets this demand is 5100 gpm.

The answer is (D).

211. The Manning coefficient for a concrete channel is 0.013. The area of the channel section is

$$A = \frac{d^2}{\tan 45°}$$
$$= d^2$$

The hydraulic radius is

$$R = \frac{d \cos 45°}{2}$$
$$= 0.353d$$

From the Manning equation, the normal depth is

$$Q = \frac{1.49}{n} A R^{2/3} \sqrt{S_o}$$
$$400 \ \frac{\text{ft}^3}{\text{sec}} = \frac{1.49}{0.013} d^2 (0.353d)^{2/3} \sqrt{0.003}$$
$$d = 6.17 \ \text{ft}$$

Since the normal depth does not equal the actual depth, the flow is not uniform. The velocity at the control structure is

$$v_1 = \frac{Q}{A_1} = \frac{400 \ \frac{\text{ft}^3}{\text{sec}}}{(5.2 \ \text{ft})^2}$$
$$= 14.79 \ \text{ft/sec}$$

The energy head at the control structure is

$$E_1 = D_1 + \frac{v_1^2}{2g}$$
$$= 5.2 \ \text{ft} + \frac{\left(14.79 \ \frac{\text{ft}}{\text{sec}}\right)^2}{(2)\left(32.2 \ \frac{\text{ft}}{\text{sec}^2}\right)}$$
$$= 8.60 \ \text{ft}$$

The hydraulic radius at the control structure is

$$R = \frac{d \cos 45°}{2} = \frac{(5.2 \ \text{ft})(0.707)}{2}$$
$$= 1.84 \ \text{ft}$$

Compute the energy head for each 0.2 ft water depth increment. The velocity for a water depth of $D_2 = 5.4$ ft is

$$v_2 = \frac{Q}{A_2} = \frac{400 \ \frac{\text{ft}^3}{\text{sec}}}{(5.4 \ \text{ft})^2}$$
$$= 13.72 \ \text{ft/sec}$$

The energy head at D_2 is

$$E_2 = D_2 + \frac{v_2^2}{2g} = 5.4 \ \text{ft} + \frac{\left(13.72 \ \frac{\text{ft}}{\text{sec}}\right)^2}{(2)\left(32.2 \ \frac{\text{ft}}{\text{sec}^2}\right)}$$
$$= 8.32 \ \text{ft}$$

The hydraulic radius at D_2 is

$$R_2 = \frac{d \cos 45°}{2} = \frac{(5.4 \ \text{ft}) \cos 45°}{2}$$
$$= 1.91 \ \text{ft}$$

The average velocity between D_1 and D_2 is

$$v_{ave} = \frac{v_1 + v_2}{2} = \frac{14.79 \ \frac{\text{ft}}{\text{sec}} + 13.72 \ \frac{\text{ft}}{\text{sec}}}{2}$$
$$= 14.26 \ \text{ft/sec}$$

The average hydraulic radius between D_1 and D_2 is

$$R_{ave} = \frac{R_1 + R_2}{2} = \frac{1.84 \ \text{ft} + 1.91 \ \text{ft}}{2}$$
$$= 1.88 \ \text{ft}$$

The average slope of the energy grade line is

$$S_{ave} = \left(\frac{n v_{ave}}{1.49 R_{ave}^{2/3}} \right)^2$$

$$= \left(\frac{(0.013) \left(14.26 \, \frac{ft}{sec} \right)}{(1.49)(1.88 \, ft)^{2/3}} \right)^2$$

$$= 0.00667$$

The length between D_1 and D_2 is

$$L = \frac{E_1 - E_2}{S_1 - S_o} = \frac{8.60 \, ft - 8.32 \, ft}{0.00667 - 0.003}$$

$$= 76.3 \, ft$$

The remaining computations are given in the accompanying table.

The depth at the highway crossing, at a distance of 345 ft upstream from the control structure, is 6.0 ft.

The answer is (C).

212. The net positive suction head available is

$$\text{NPSHA} = h_{atm} + h_{z(s)} - h_{f(s)} - h_{vp}$$

The atmospheric pressure at 5000 ft altitude is 12 psi. The atmospheric head is

$$h_{atm} = \frac{p}{\gamma} = \frac{\left(12 \, \frac{lbf}{in^2} \right) \left(144 \, \frac{in^2}{ft^2} \right)}{62.4 \, \frac{lbf}{ft^3}}$$

$$= 27.7 \, ft$$

The vapor pressure of water at 90°F is 1.62 ft. The friction loss in the suction piping is 3.6 ft. The NPSHA is

$$\text{NPSHA} = 1.3 \text{NPSHR}$$

$$= (1.3)(10.3 \, ft)$$

$$= 13.39 \, ft$$

The elevation difference is

$$h_{z(s)} = \text{NPSHA} - h_{atm} + h_{f(s)} + h_{vp}$$

$$= 13.39 \, ft - 27.7 \, ft + 3.6 \, ft + 1.62 \, ft$$

$$= -9.09 \, ft \quad (-9.0 \, ft)$$

The pump centerline must be no more than 9 ft above the water level in the pit.

The answer is (B).

213. For a triangular channel, the area is

$$A = \tfrac{1}{2} bh = \left(\tfrac{1}{2} \right) \left(\sqrt{2} d \right)^2 = d^2$$

The velocity is

$$v = \frac{Q}{A}$$

Since $A = d^2$,

$$v = \frac{Q}{d^2} = \frac{200 \, \frac{ft^3}{sec}}{(3.2 \, ft)^2}$$

$$= 19.53 \, ft/sec$$

The hydraulic depth, D_h, is the ratio of the area in flow to the width of the channel at the fluid surface. Since $A = d_2$ and the width for a 90° internal angle is $2d$,

$$D_h = \frac{d^2}{2d} = \frac{(3.2 \, ft)^2}{(2)(3.2 \, ft)}$$

$$= 1.6 \, ft$$

Table for Solution 211

depth, D (ft)	velocity, v (ft/sec)	head, $v^2/2g$ (ft)	energy, E (ft)	hydraulic radius, R (ft)	average velocity, v_{ave} (ft/sec)	average hydraulic radius, R_{ave} (ft)	average slope, S_{ave} (ft/ft)	change in energy, ΔE (ft)	length, L (ft)	cumulative length, L_{cum} (ft)
5.2	14.79	3.40	8.60	1.84	–	–	–	–	–	0
–	–	–	–	–	14.26	1.88	0.00667	0.28	76.3	–
5.4	13.72	2.92	8.32	1.91	–	–	–	–	–	76
–	–	–	–	–	13.24	1.95	0.00545	0.19	78	–
5.6	12.76	2.53	8.13	1.98	–	–	–	–	–	154
–	–	–	–	–	12.33	2.02	0.00451	0.13	86	–
5.8	11.89	2.20	8.00	2.05	–	–	–	–	–	240
–	–	–	–	–	11.50	2.09	0.00375	0.08	106	–
6.0	11.11	1.92	7.92	2.12	–	–	–	–	–	346

The Froude number is

$$Fr = \frac{v}{\sqrt{gD_h}}$$

$$= \frac{19.53 \ \dfrac{ft}{sec}}{\sqrt{\left(32.2 \ \dfrac{ft}{sec^2}\right)(1.6 \ ft)}}$$

$$= 2.72$$

The Froude number is greater than 1, so the flow is supercritical.

The answer is (C).

214. The area of the rectangular channel is

$$A = dw = d_1 w$$
$$= d_1(9 \ ft)$$

The hydraulic radius is

$$R = \frac{A}{P} = \frac{dw}{2d + w}$$
$$= \frac{d_1(9 \ ft)}{2d_1 + 9 \ ft}$$

The normal depth is

$$Q = \frac{1.49}{n} AR^{2/3}\sqrt{S}$$

$$530 \ \frac{ft^3}{sec} = \left(\frac{1.49}{0.01}\right)(9 \ ft)d_1$$

$$\times \left(\frac{d_1(9 \ ft)}{2d_1 + 9 \ ft}\right)^{2/3}\sqrt{0.004}$$

$$d_1 = 3.84 \ ft$$

The upstream velocity is

$$v_1 = \frac{Q}{A} = \frac{530 \ \dfrac{ft^3}{sec}}{(9 \ ft)(3.84 \ ft)}$$

$$= 15.34 \ ft/sec$$

The Froude number for a rectangular channel is

$$Fr = \frac{v}{\sqrt{gd_1}}$$

$$= \frac{15.34 \ \dfrac{ft}{sec}}{\sqrt{\left(32.2 \ \dfrac{ft}{sec^2}\right)(3.84 \ ft)}}$$

$$= 1.379$$

The Froude number is greater than 1; therefore, the flow is supercritical.

The downstream depth is

$$\frac{d_2}{d_1} = \frac{1}{2}\left(\sqrt{1 + 8(Fr)^2} - 1\right)$$

$$d_2 = \left(\frac{3.84 \ ft}{2}\right)\left(\sqrt{1 + (8)(1.379)^2} - 1\right)$$

$$= 5.81 \ ft$$

The downstream flow is subcritical.

The energy loss is

$$\Delta E = \frac{(d_2 - d_1)^3}{4d_1 d_2}$$

$$= \frac{(5.81 \ ft - 3.84 \ ft)^3}{(4)(3.84 \ ft)(5.81 \ ft)}$$

$$= 0.086 \ ft \quad (0.09 \ ft)$$

The answer is (A).

215. The average discharge from the tank is

$$\dot{V}_{ave} = \frac{V}{t} = \frac{6.66 \ ft^3}{2 \ sec}$$

$$= 3.33 \ ft^3/sec$$

The peak discharge from the tank is

$$\dot{V}_{peak} = \left(3.33 \ \frac{ft^3}{sec}\right)\left(\frac{150\%}{100\%}\right)$$

$$= 5 \ ft^3/sec$$

The area of the throat is

$$A = \frac{\pi D^2}{4} = \frac{\pi(6 \ in)^2 \left(\dfrac{1 \ ft}{12 \ in}\right)^2}{4}$$

$$= 0.196 \ ft^2$$

The density of mercury is 848.6 lbm/ft^3. The density of water at 50°F is 62.41 lbm/ft^3. The flow coefficient is

$$C_f = \frac{A}{\dot{V}_{peak}}\sqrt{\frac{2g(\rho_m - \rho_w)h}{\rho_w}}$$

$$= \left(\frac{0.196 \ ft^2}{5 \ \dfrac{ft^3}{sec}}\right)$$

$$\times \sqrt{\frac{\begin{matrix}(2)\left(32.2 \ \dfrac{ft}{sec^2}\right)\left(848.6 \ \dfrac{lbm}{ft^3} - 62.41 \ \dfrac{lbm}{ft^3}\right) \\ \times (8 \ in)\left(\dfrac{1 \ ft}{12 \ in}\right)\end{matrix}}{62.41 \ \dfrac{lbm}{ft^3}}}$$

$$= 0.911 \quad (0.91)$$

The answer is (C).

216. The friction loss in the pipe is

$$h_f = f\left(\frac{L}{D}\right)\left(\frac{v^2}{2g}\right)$$

Substitute the given head loss of 1.5 m/km and the pipe dimensions into the friction loss equation.

$$1.5 \text{ m} = f\left(\frac{(1 \text{ km})\left(1000 \frac{\text{m}}{\text{km}}\right)}{2.5 \text{ m}}\right)\left(\frac{v^2}{(2)\left(9.81 \frac{\text{m}}{\text{s}^2}\right)}\right)$$

$$fv^2 = 0.073\,58$$

$$v = \sqrt{\frac{0.073\,58 \text{ m}^2}{f}} \qquad \text{[I]}$$

For concrete pipe, $\epsilon = 1.2 \times 10^{-3}$ m. The relative roughness is

$$\frac{\epsilon}{D} = \frac{1.2 \times 10^{-3} \text{ m}}{2.5 \text{ m}}$$
$$= 0.000\,48$$

At 10°C, the kinematic viscosity of water is 1.31×10^{-6} m²/s.

The Reynolds number is

$$\text{Re} = \frac{Dv}{\nu} = \frac{(2.5 \text{ m})v}{1.31 \times 10^{-6} \frac{\text{m}^2}{\text{s}}}$$
$$= \left(1.908 \times 10^6 \frac{\text{s}}{\text{m}}\right)v \qquad \text{[II]}$$

Solve by iteration by substituting an initial estimate of $f = 0.02$ into Eq. I.

$$v^2 = \frac{0.073\,58 \text{ m}^2}{0.02}$$
$$v = 1.918 \text{ m/s}$$

Substituting into Eq. II,

$$\text{Re} = \left(1.908 \times 10^6 \frac{\text{s}}{\text{m}}\right)\left(1.918 \frac{\text{m}}{\text{s}}\right)$$
$$= 3.66 \times 10^6$$

From a Moody diagram, the friction factor is

$$f = 0.017$$

Perform a second trial with a friction factor of 0.017, using Eq. I.

$$v = \sqrt{\frac{0.073\,58 \text{ m}^2}{0.017}}$$
$$= 2.080 \text{ m/s}$$

The revised Reynolds number is

$$\text{Re} = \left(1.908 \times 10^6 \frac{\text{s}}{\text{m}}\right)\left(2.08 \frac{\text{m}}{\text{s}}\right)$$
$$= 3.969 \times 10^6$$

From a Moody diagram,

$$f = 0.0175$$

This friction factor matches closely the friction factor used in the second trial. Therefore,

$$v = 2.08 \text{ m/s}$$

The discharge is

$$Q = vA = v\frac{\pi D^2}{4}$$
$$= \left(2.08 \frac{\text{m}}{\text{s}}\right)\pi\frac{(2.5 \text{ m})^2}{4}$$
$$= 10.2 \text{ m}^3/\text{s} \quad (10 \text{ m}^3/\text{s})$$

The answer is (A).

217. From a diagram of hydraulic elements of circular sections for a Manning roughness constant of $n = 0.012$, the optimum discharge occurs with the depth of flow-to-diameter ratio, d/D, of 0.93. At this ratio, the discharge ratio is

$$\frac{Q}{Q_f} = 1.075$$

The optimum depth is 0.93 of the diameter of the pipe. When critical velocity occurs at optimum depth, the discharge is a maximum for the available energy. Optimum discharge is, therefore, obtained by making the critical depth the same as the optimum depth. For a 24 in pipe, the critical depth is

$$d_c = 0.93D$$
$$= (0.93)(24 \text{ in})\left(\frac{1 \text{ ft}}{12 \text{ in}}\right)$$
$$= 1.86 \text{ ft}$$

From the hydraulic elements diagram for $d/D = 0.93$,

$$\frac{A}{A_f} = 0.96$$
$$\frac{R}{R_f} = 1.17$$

The area is

$$A = (0.96)\frac{\pi(24 \text{ in})^2}{4}\left(\frac{1 \text{ ft}}{12 \text{ in}}\right)^2$$
$$= 3.02 \text{ ft}^2$$

The full hydraulic radius is

$$R_{\text{full}} = \frac{A_{\text{full}}}{P_{\text{full}}} = \frac{\pi D^2}{4\pi D}$$
$$= \frac{D}{4}$$
$$= \frac{2 \text{ ft}}{4}$$
$$= 0.5 \text{ ft}$$

The hydraulic radius at critical depth is

$$\frac{R}{R_{\text{full}}} = 1.17$$
$$R = (0.5 \text{ ft})(1.17)$$
$$= 0.585 \text{ ft}$$

The critical slope can be found from the Manning equation.

$$S = \left(\frac{Qn}{1.486AR^{2/3}}\right)^2$$
$$= \left(\frac{\left(30 \frac{\text{ft}^3}{\text{sec}}\right)(0.012)}{(1.486)(3.02 \text{ ft}^2)(0.585 \text{ ft})^{2/3}}\right)^2$$
$$= 0.013 \text{ ft/ft}$$

The answer is (B).

218. The total upstream head is

$$H_o = h_o + \frac{v_o^2}{2g} = 6 \text{ ft} + \frac{\left(10 \frac{\text{ft}}{\text{sec}}\right)^2}{(2)\left(32.2 \frac{\text{ft}}{\text{sec}^2}\right)}$$
$$= 7.55 \text{ ft}$$

The discharge is

$$Q = bac\sqrt{2g}\frac{H_o}{\sqrt{H_o + \Psi a}}$$
$$= (2 \text{ ft})(2 \text{ ft})(0.603)\sqrt{(2)\left(32.2 \frac{\text{ft}}{\text{sec}^2}\right)}$$
$$\times \left(\frac{7.55 \text{ ft}}{\sqrt{7.55 \text{ ft} + (0.624)(2 \text{ ft})}}\right)$$
$$= 49.27 \text{ ft}^3/\text{sec}$$

The depth of flow at the *vena contracta* (jet) is given by

$$H_1 = \Psi a = (0.624)(2 \text{ ft})$$
$$= 1.248 \text{ ft}$$

The velocity is

$$v = \frac{Q}{A} = \frac{\left(49.27 \frac{\text{ft}^3}{\text{sec}}\right)}{(1.248 \text{ ft})(2 \text{ ft})}$$
$$= 19.74 \text{ ft/sec} \quad (20 \text{ ft/sec})$$

The answer is (D).

219. The density of water at 60°F is 62.37 lbm/ft^3. The density of gasoline is

$$\rho_{\text{gas}} = \text{SG}\rho_{\text{water}}$$
$$= (0.80)\left(62.37 \frac{\text{lbm}}{\text{ft}^3}\right)$$
$$= 49.90 \text{ lbm/ft}^3$$

The compressibility of gasoline at 60°F is

$$\beta = \frac{1.0 \times 10^{-5}}{\text{psi}}$$

The bulk modulus is the reciprocal of compressibility, and for gasoline it is

$$E_{\text{gas}} = \frac{1}{\beta} = \frac{1}{\dfrac{1.0 \times 10^{-5}}{\text{psi}}}$$
$$= 1 \times 10^5 \text{ psi}$$

The effective bulk modulus of elasticity of the pipe and the gasoline is

$$E = \frac{E_{\text{gas}}t_{\text{pipe}}E_{\text{pipe}}}{t_{\text{pipe}}E_{\text{pipe}} + D_{\text{pipe}}E_{\text{gas}}}$$

$$= \frac{\begin{pmatrix}\left(1 \times 10^5 \frac{\text{lbf}}{\text{in}^2}\right)(0.25 \text{ in})\left(\frac{1 \text{ ft}}{12 \text{ in}}\right) \\ \times \left(29 \times 10^6 \frac{\text{lbf}}{\text{in}^2}\right)\end{pmatrix}}{\begin{pmatrix}(0.25 \text{ in})\left(\frac{1 \text{ ft}}{12 \text{ in}}\right)\left(29 \times 10^6 \frac{\text{lbf}}{\text{in}^2}\right) \\ + (8 \text{ in})\left(\frac{1 \text{ ft}}{12 \text{ in}}\right)\left(1 \times 10^5 \frac{\text{lbf}}{\text{in}^2}\right)\end{pmatrix}}$$

$$= 90.06 \times 10^3 \text{ lbf/in}^2$$

The speed of sound in the pipe is

$$a = \sqrt{\frac{Eg_c}{\rho}}$$

$$= \sqrt{\frac{\left(90.06 \times 10^3 \; \frac{lbf}{in^2}\right)\left(12 \; \frac{in}{ft}\right)^2 \left(32.2 \; \frac{ft\text{-}lbm}{lbf\text{-}sec^2}\right)}{49.90 \; \frac{lbm}{ft^3}}}$$

$$= 2893 \; ft/sec$$

The pressure increase is

$$\Delta p = \frac{\rho a \Delta v}{g_c} = \frac{\left(49.90 \; \frac{lbm}{ft^3}\right)\left(2893 \; \frac{ft}{sec}\right)\left(10 \; \frac{ft}{sec}\right)}{\left(32.2 \; \frac{ft\text{-}lbm}{lbf\text{-}sec^2}\right)\left(144 \; \frac{in^2}{ft^2}\right)}$$

$$= 311.3 \; lbf/in^2$$

Check that the closure is rapid for full development of the pressure surge. The maximum time for the pressure wave to return is

$$t = \frac{2L}{a} = \frac{(2)(2500 \; ft)}{2983 \; \frac{ft}{sec}}$$

$$= 1.676 \; sec$$

The valve closes before the wave returns, so the maximum pressure of 311.3 lbf/in^2 (310 psi) is developed.

The answer is (D).

220. Since depth and velocity are inversely proportional, the velocity can be scaled down from the point of minimum depth.

$$v_o d_o = v_1 d_1$$

$$\left(62 \; \frac{ft}{sec}\right)(4.08 \; ft) = v_1 (4.57 \; ft)$$

$$v_1 = 55.35 \; ft/sec$$

The depths immediately before and after a hydraulic jump are the conjugate depths, d_1 and d_2. One can be calculated from the other.

$$d_2 = -\frac{d_1}{2} + \sqrt{\frac{2v_1^2 d_1}{g} + \frac{d_1^2}{4}}$$

$$= -\frac{4.57 \; ft}{2}$$

$$+ \sqrt{\frac{(2)\left(55.35 \; \frac{ft}{sec}\right)^2 (4.57 \; ft)}{32.2 \; \frac{ft}{sec^2}} + \frac{(4.57 \; ft)^2}{4}}$$

$$= 27.29 \; ft \quad (27 \; ft)$$

The answer is (D).

221. Use a double-mass analysis to test the consistency of the record at station C with the group of surrounding stations. Plot the accumulated annual precipitation of station C against the concurrent accumulated annual precipitation of the surrounding stations as shown in the following table and graph.

	station C		adjacent 10 stations	
	annual	cumulative	annual	cumulative
	precipitation	precipitation	precipitation	precipitation
year	(in)	(in)	(in)	(in)
1950	–	0	–	0
1951	25.0	25.0	30.0	30.0
1952	24.6	49.6	31.2	61.2
1953	26.3	75.9	29.7	90.9
1954	25.7	101.6	30.5	121.4
1955	24.3	125.9	28.7	150.1
1956	25.9	151.8	31.3	181.4
1957	27.3	179.1	30.6	212.0
1958	24.5	203.6	29.9	241.9
1959	23.9	227.5	29.8	271.7
1960	22.5	250.0	28.3	300.0
1961	30.0	280.0	20.0	320.0
1962	31.2	311.2	21.2	341.2
1963	29.8	341.0	19.9	361.1
1964	30.5	371.5	20.5	381.6
1965	28.5	400.0	18.4	400.0

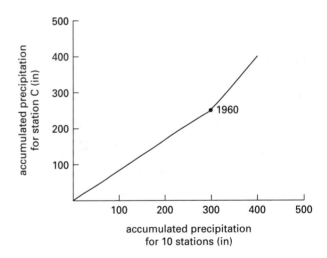

The curve connects points that represent the data years. There is a significant break in the slope of the accumulated precipitation after 1960 at station C compared to the 10 stations. The data from station C cannot be used to characterize the storm without adjustment.

The answer is (B).

222. The normal annual precipitation of station D relative to stations A, B, and C is

$$\frac{\text{station A}}{\text{station D}} = \frac{40.5 \text{ in}}{50.3 \text{ in}}$$
$$= 0.805$$
$$\frac{\text{station B}}{\text{station D}} = \frac{32.8 \text{ in}}{50.3 \text{ in}}$$
$$= 0.652$$
$$\frac{\text{station C}}{\text{station D}} = \frac{60.7 \text{ in}}{50.3 \text{ in}}$$
$$= 1.207$$

The normal annual precipitation of station D, which is missing a value for Storm Alpha precipitation, varies by more than 10% from the other stations. The normal-ratio method, which applies if the precipitation difference between locations is more than 10%, can be used to estimate the unknown precipitation at station D. The missing Storm Alpha precipitation is

$$P_x = \frac{1}{3}\left(\left(\frac{N_x}{N_A}\right)P_A + \left(\frac{N_x}{N_B}\right)P_B + \left(\frac{N_x}{N_C}\right)P_C\right)$$

$$= \left(\frac{1}{3}\right)\begin{pmatrix} \left(\dfrac{50.3 \text{ in}}{40.5 \text{ in}}\right)(3.20 \text{ in}) \\ + \left(\dfrac{50.3 \text{ in}}{32.8 \text{ in}}\right)(2.70 \text{ in}) \\ + \left(\dfrac{50.3 \text{ in}}{60.7 \text{ in}}\right)(3.90 \text{ in}) \end{pmatrix}$$

$$= 3.78 \text{ in}$$

The average precipitation over the watershed is

$$P_{\text{avg}} = \frac{P_A A_A + P_B A_B + P_C A_C + P_D A_D}{A_A + A_B + A_C + A_D}$$

$$= \frac{\begin{pmatrix} (3.20 \text{ in})(2000 \text{ ac}) \\ + (2.70 \text{ in})(2500 \text{ ac}) \\ + (3.90 \text{ in})(2800 \text{ ac}) \\ + (3.78 \text{ in})(2400 \text{ ac}) \end{pmatrix}}{2000 \text{ ac} + 2500 \text{ ac} + 2800 \text{ ac} + 2400 \text{ ac}}$$

$$= 3.41 \text{ in}$$

The rainfall intensity is

$$I = \frac{P_{\text{avg}}}{t} = \left(\frac{3.41 \text{ in}}{57 \text{ min}}\right)\left(60 \frac{\text{min}}{\text{hr}}\right)$$
$$= 3.59 \text{ in/hr}$$

From the given intensity-duration-frequency curve, the return frequency is most nearly 50 yr.

The answer is (D).

223. Statement II is false. The time of concentration is not used to construct a unit hydrograph.

Statement V is false. The rational method is not used for durations of storms that are multiples of the storm duration for which a unit hydrograph is available.

Statements I, III, and IV are true.

The answer is (B).

224. The Doorenbos and Pruitt vapor transport coefficient is

$$B = 0.0027\left(1 + \frac{u}{100}\right)$$

u is the 24 hr wind run in km/d.

$$B = (0.0027)\left(1 + \frac{260 \dfrac{\text{km}}{\text{d}}}{100}\right)$$

$$= 0.009\,72 \text{ mm/d·Pa}$$

The saturated vapor pressure of water vapor over liquid water at 20°C, e_{as}, is 2.338 kPa, or 2338 Pa. The relative humidity, R_h, is 30%, or 0.30.

$$e_a = R_h e_{\text{as}}$$
$$= (0.30)(2338 \text{ Pa})$$
$$= 701.4 \text{ Pa}$$

The evaporation rate from aerodynamic vapor transport is

$$E_a = B(e_{\text{as}} - e_a)$$
$$= \left(0.009\,72 \frac{\text{mm}}{\text{Pa·d}}\right)(2338 \text{ Pa} - 701.4 \text{ Pa})$$
$$= 15.91 \text{ mm/d}$$

The latent heat of vaporization is

$$l_v = 2500 - 2.36T$$
$$= 2500 - (2.36)(20°\text{C})$$
$$= 2452.8 \text{ kJ/kg}$$

The density of water at 20°C is 998.23 kg/m³. The evaporation rate from radiation is

$$E_r = \frac{R_n}{l_v \rho_w}$$

$$= \frac{\left(250 \dfrac{\text{W}}{\text{m}^2}\right)\left(10^3 \dfrac{\text{mm}}{\text{m}}\right)\left(86\,400 \dfrac{\text{s}}{\text{d}}\right)}{\left(2452.8 \dfrac{\text{kJ}}{\text{kg}}\right)\left(1000 \dfrac{\text{J}}{\text{kJ}}\right)\left(998.23 \dfrac{\text{kg}}{\text{m}^3}\right)}$$

$$= 8.82 \text{ mm/d}$$

The evapotranspiration for the reference crop is the sum of the radiation evaporation rate, multiplied by its weighting factor, and the vapor transport evaporation rate, multiplied by its weighting factor.

$$E_{\text{tr}} = \frac{\Delta}{\Delta + \gamma} E_r + \frac{\gamma}{\Delta + \gamma} E_a$$
$$= (0.7)\left(8.82 \ \frac{\text{mm}}{\text{d}}\right) + (0.3)\left(15.91 \ \frac{\text{mm}}{\text{d}}\right)$$
$$= 10.95 \ \text{mm/d}$$

The evapotranspiration for the alfalfa is

$$E_t = k_s k_c E_{\text{tr}}$$
$$= (0.80)(0.35)\left(10.95 \ \frac{\text{mm}}{\text{d}}\right)$$
$$= 3.07 \ \text{mm/d} \quad (3 \ \text{mm/d})$$

The answer is (B).

225. The cumulative infiltration depth is

$$F = \frac{V}{A} = \frac{12 \ \text{in}^3}{6 \ \text{in}^2}$$
$$= 2 \ \text{in}$$

The cumulative infiltration depth is a function of the soil suction potential and the hydraulic conductivity. The sorptivity is

$$S = \frac{F}{\sqrt{t}} = \frac{2 \ \text{in}}{\sqrt{\dfrac{(20 \ \text{min})(1 \ \text{hr})}{60 \ \text{min}}}}$$
$$= 3.464 \ \text{in}/\sqrt{\text{hr}}$$

The cumulative infiltration depth is

$$F = S\sqrt{t} + Kt$$
$$= \left(3.464 \ \frac{\text{in}}{\sqrt{\text{hr}}}\right)\left(\sqrt{1 \ \text{hr}}\right) + \left(1 \ \frac{\text{in}}{\text{hr}}\right)(1 \ \text{hr})$$
$$= 4.46 \ \text{in} \quad (4.0 \ \text{in})$$

The answer is (B).

226. Determine the intensity of precipitation for each subduration and arrange them in descending order. The intensity for the 1.9 cm rainfall depth for 5 min duration is

$$I = \left(\frac{1.9 \ \text{cm}}{5 \ \text{min}}\right)\left(\frac{60 \ \text{min}}{1 \ \text{h}}\right)$$
$$= 22.8 \ \text{cm/h}$$

The probability is

$$P = \frac{m}{N + 1 \ \text{yr}}$$

m is the order number, and N is the number of years.

$$P = \frac{m}{10 \ \text{yr} + 1 \ \text{yr}}$$
$$= \frac{m}{11 \ \text{yr}}$$

The calculations for the 5 min and 20 min durations are given in the following table.

5 min duration

depth (cm)	intensity (cm/h)	order number, m	probability
1.9	22.8	1	0.09
1.8	21.6	2	0.18
1.5	18.0	3	0.27
1.3	15.6	4	0.36
1.2	14.4	5	0.45
1.2	14.4	6	0.55
1.0	12.0	7	0.64
0.9	10.8	8	0.73
0.8	9.6	9	0.82
0.7	8.4	10	0.91

20 min duration

depth (cm)	intensity (cm/h)	order number, m	probability
3.2	9.6	1	0.09
3.1	9.3	2	0.18
2.9	8.7	3	0.27
2.8	8.4	4	0.36
2.6	7.8	5	0.45
2.5	7.5	6	0.55
2.4	7.2	7	0.64
2.4	7.2	8	0.73
2.3	6.9	9	0.82
2.1	6.3	10	0.91

Determine the probability of the 2-yr and 10-yr return period.

$$P = \frac{1}{T}$$
$$P_{\text{2-yr}} = \frac{1}{2 \ \text{yr}} = 0.5$$
$$P_{\text{10-yr}} = \frac{1}{10 \ \text{yr}} = 0.1$$

The intensity for the 2-yr ($P_{2\text{-yr}} = 0.5$) return period for the 5 min subduration is found by interpolation in the table.

$$x = 14.4 \ \frac{\text{cm}}{\text{h}} - \left(14.4 \ \frac{\text{cm}}{\text{h}} - 14.4 \ \frac{\text{cm}}{\text{h}} \right) \left(\frac{0.5 - 0.45}{0.55 - 0.45} \right)$$
$$= 14.4 \ \text{cm/h}$$

The intensity for the 10-yr ($P_{10\text{-yr}} = 0.1$) return period for the 5 min subduration is found by interpolation.

$$x = 22.8 \ \frac{\text{cm}}{\text{h}} - \left(22.8 \ \frac{\text{cm}}{\text{h}} - 21.6 \ \frac{\text{cm}}{\text{h}} \right) \left(\frac{0.1 - 0.09}{0.18 - 0.09} \right)$$
$$= 22.67 \ \text{cm/h}$$

The intensity for the 2-yr ($P_{2\text{-yr}} = 0.5$) return period for the 20 min subduration is found by interpolation.

$$x = 7.8 \ \frac{\text{cm}}{\text{h}} - \left(7.8 \ \frac{\text{cm}}{\text{h}} - 7.5 \ \frac{\text{cm}}{\text{h}} \right) \left(\frac{0.5 - 0.45}{0.55 - 0.45} \right)$$
$$= 7.65 \ \text{cm/h}$$

The intensity for the 10-yr ($P_{10\text{-yr}} = 0.1$) return period for the 20 min subduration is found by interpolation.

$$x = 9.6 \ \frac{\text{cm}}{\text{h}} - \left(9.6 \ \frac{\text{cm}}{\text{h}} - 9.3 \ \frac{\text{cm}}{\text{h}} \right) \left(\frac{0.1 - 0.09}{0.18 - 0.09} \right)$$
$$= 9.57 \ \text{cm/h}$$

The intensities for the required return periods for each subduration are given in the following table.

subduration	2-yr return intensity (cm/h)	10-yr return intensity (cm/h)
5 min	14.4	22.67
20 min	7.65	9.57

For a 15 min storm, the 2-yr intensity for the 2-yr period is

$$x = 14.40 \ \frac{\text{cm}}{\text{h}} - \left(14.4 \ \frac{\text{cm}}{\text{h}} - 7.65 \ \frac{\text{cm}}{\text{h}} \right)$$
$$\times \left(\frac{15 \ \text{min} - 5 \ \text{min}}{20 \ \text{min} - 5 \ \text{min}} \right)$$
$$= 9.90 \ \text{cm/h}$$

For the 10-yr return period,

$$x = 22.67 \ \frac{\text{cm}}{\text{h}} - \left(22.67 \ \frac{\text{cm}}{\text{h}} - 9.57 \ \frac{\text{cm}}{\text{h}} \right)$$
$$\times \left(\frac{15 \ \text{min} - 5 \ \text{min}}{20 \ \text{min} - 5 \ \text{min}} \right)$$
$$= 13.94 \ \text{cm/h}$$

The percent increase is

$$\left(\frac{13.94 \ \frac{\text{cm}}{\text{h}} - 9.90 \ \frac{\text{cm}}{\text{h}}}{9.90 \ \frac{\text{cm}}{\text{h}}} \right) \times 100\% = 40.8\% \quad (40\%)$$

The answer is (D).

227. Sheet flow depth and unit width are shown in the following illustration.

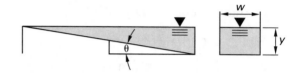

The intensity is

$$i = \left(1.3 \ \frac{\text{in}}{\text{hr}} \right) \left(\frac{1 \ \text{ft}}{12 \ \text{in}} \right) \left(\frac{\text{hr}}{3600 \ \text{sec}} \right)$$
$$= 30 \times 10^{-6} \ \text{ft/sec}$$

The slope angle θ is

$$\theta = \tan^{-1} S_o = \tan^{-1} 0.06$$
$$= 3.43°$$

The kinematic viscosity at 60°F is $1.217 \times 10^{-5} \ \text{ft}^2/\text{sec}$. The discharge per unit width is

$$q_o = \text{v}y$$
$$= (i - f)L_o \cos \theta$$
$$= \left(30 \times 10^{-6} \ \frac{\text{ft}}{\text{sec}} - 0 \right) (150 \ \text{ft}) \cos 3.43°$$
$$= 4.5 \times 10^{-3} \ \text{ft}^3/\text{sec·ft}$$

For sheet flow with unit width, the hydraulic radius is

$$R = \frac{A}{P} = \frac{wy}{w}$$
$$= y$$

The Reynolds number is

$$\text{Re} = \frac{D_e \text{v}}{\nu} = \frac{4R\text{v}}{\nu}$$
$$= \frac{4y\text{v}}{\nu}$$
$$= \frac{4q_o}{\nu}$$
$$= \frac{(4) \left(4.5 \times 10^{-3} \ \frac{\text{ft}^3}{\text{sec·ft}} \right)}{1.217 \times 10^{-5} \ \frac{\text{ft}^2}{\text{sec}}}$$
$$= 1479$$

The Reynolds number is less than 2000, so the flow is laminar. For laminar flow under rainfall, the friction factor increases with rainfall intensity according to the following equations.

$$C_L = 96 + 108i^{0.4}$$
$$= 96 + (108)\left(1.3 \; \frac{\text{in}}{\text{hr}}\right)^{0.4}$$
$$= 216$$

The friction factor is

$$f = \frac{C_L}{\text{Re}} = \frac{216}{1479}$$
$$= 0.146$$

The depth of sheet flow is

$$y = \left(\frac{fq_o^2}{8gS_o}\right)^{1/3}$$
$$= \left(\frac{(0.146)\left(4.5 \times 10^{-3} \; \frac{\text{ft}^3}{\text{sec·ft}}\right)^2}{(8)\left(32.2 \; \frac{\text{ft}}{\text{sec}^2}\right)(0.06)}\right)^{1/3}$$
$$= 0.00576 \text{ ft}$$

The velocity is

$$\text{v} = \frac{q_o}{y} = \frac{4.5 \times 10^{-3} \; \frac{\text{ft}^3}{\text{sec·ft}}}{0.00576 \text{ ft}}$$
$$= 0.781 \text{ ft/sec} \quad (0.80 \text{ ft/sec})$$

The answer is (B).

228. The precipitation at the proposed development area is

$$P_x = \frac{d_{A-x}^2 P_A + d_{B-x}^2 P_B + d_{C-x}^2 P_C + d_{D-x}^2 P_D}{d_{A-x}^2 + d_{B-x}^2 + d_{C-x}^2 + d_{D-x}^2}$$

$$= \frac{\begin{array}{c}(6.3 \text{ mi})^2 \left(3.2 \; \frac{\text{in}}{\text{hr}}\right) + (4.2 \text{ mi})^2 \left(4.6 \; \frac{\text{in}}{\text{hr}}\right) \\ + (5.1 \text{ mi})^2 \left(2.9 \; \frac{\text{in}}{\text{hr}}\right) + (8.1 \text{ mi})^2 \left(3.7 \; \frac{\text{in}}{\text{hr}}\right)\end{array}}{(6.3 \text{ mi})^2 + (4.2 \text{ mi})^2 + (5.1 \text{ mi})^2 + (8.1 \text{ mi})^2}$$

$$= 3.53 \text{ in/hr} \quad (3.5 \text{ in/hr})$$

The answer is (B).

229. At steady flow, the discharge can be determined by summing the flow in each section of the channel and flood plain. Consider the sections as parallel flows with no wetted perimeter between the interface. From a table of Manning roughness coefficients, the coefficients for each section are

channel	0.025
west plain	0.035
east plain	0.060

For the channel, the area is

$$A_{\text{channel}} = d_{\text{channel}} w_{\text{channel}}$$
$$= (8 \text{ ft})(50 \text{ ft})$$
$$= 400 \text{ ft}^2$$

The wetted perimeter is

$$P_{\text{channel}} = (d_{\text{channel}} - d_{\text{west}}) + w_{\text{channel}}$$
$$+ (d_{\text{channel}} - d_{\text{east}})$$
$$= (8 \text{ ft} - 2 \text{ ft}) + 50 \text{ ft} + (8 \text{ ft} - 1 \text{ ft})$$
$$= 63 \text{ ft}$$

The hydraulic radius is

$$R_{\text{channel}} = \frac{A_{\text{channel}}}{P_{\text{channel}}} = \frac{400 \text{ ft}^2}{63 \text{ ft}}$$
$$= 6.35 \text{ ft}$$

The discharge is

$$Q_{\text{channel}} = \frac{1.49}{n} A_{\text{channel}} R_{\text{channel}}^{2/3} \sqrt{S_o}$$
$$= \left(\frac{1.49}{0.025}\right)(400 \text{ ft}^2)(6.35 \text{ ft})^{2/3}\sqrt{0.002}$$
$$= 3656 \text{ ft}^3/\text{sec}$$

For the west flood plain, the area is

$$A_{\text{west}} = (2 \text{ ft})(300 \text{ ft})$$
$$= 600 \text{ ft}^2$$

The wetted perimeter is

$$P_{\text{west}} = d_{\text{west}} + w_{\text{west}}$$
$$= 2 \text{ ft} + 300 \text{ ft}$$
$$= 302 \text{ ft}$$

The hydraulic radius is

$$R_{\text{west}} = \frac{600 \text{ ft}^2}{302 \text{ ft}}$$
$$= 1.99 \text{ ft}$$

The discharge is

$$Q_{\text{west}} = \left(\frac{1.49}{0.035}\right)(600 \text{ ft}^2)(1.99 \text{ ft})^{2/3}\sqrt{0.002}$$
$$= 1807.3 \text{ ft}^3/\text{sec}$$

For the east flood plain, the area is

$$A_{\text{east}} = (1 \text{ ft})(400 \text{ ft})$$
$$= 400 \text{ ft}^2$$

The wetted perimeter is

$$P_{\text{east}} = d_{\text{east}} + w_{\text{east}}$$
$$= 1 \text{ ft} + 400 \text{ ft}$$
$$= 401 \text{ ft}$$

The hydraulic radius is

$$R_{\text{east}} = \frac{400 \text{ ft}^2}{401 \text{ ft}}$$
$$= 0.99 \text{ ft}$$

The discharge is

$$Q_{\text{east}} = \left(\frac{1.49}{0.060}\right)(400 \text{ ft}^2)(0.99 \text{ ft})^{2/3}\sqrt{0.002}$$
$$= 441.3 \text{ ft}^3/\text{sec}$$

The total flood flow is

$$Q_{\text{total}} = Q_{\text{channel}} + Q_{\text{west}} + Q_{\text{east}}$$
$$= 3656 \frac{\text{ft}^3}{\text{sec}} + 1807.3 \frac{\text{ft}^3}{\text{sec}} + 441.3 \frac{\text{ft}^3}{\text{sec}}$$
$$= 5905 \text{ ft}^3/\text{sec} \quad (6000 \text{ ft}^3/\text{sec})$$

The answer is (C).

230. Assume Stokes' law applies. The settling velocity is

$$v_{s,\text{actual}} = \frac{D^2 g(\rho_{\text{particle}} - \rho_{\text{water}})}{18\mu g_c}$$
$$= \frac{D^2 g(SG_{\text{particle}} - SG_{\text{water}})}{18\nu}$$
$$= \frac{(4 \times 10^{-4} \text{ in})^2 \left(32.2 \frac{\text{ft}}{\text{sec}^2}\right)}{\times (2.65 - 1)\left(3600 \frac{\text{sec}}{\text{hr}}\right)}$$
$$= \frac{}{(18)\left(1.664 \times 10^{-5} \frac{\text{ft}^2}{\text{sec}}\right)\left(12 \frac{\text{in}}{\text{ft}}\right)^2}$$
$$= 0.71 \text{ ft/hr}$$

Check that the Reynolds number is less than 1.

$$Re = \frac{v_s D}{\nu}$$
$$= \frac{\left(0.71 \frac{\text{ft}}{\text{hr}}\right)(4 \times 10^{-4} \text{ in})}{\left(1.664 \times 10^{-5} \frac{\text{ft}^2}{\text{sec}}\right)\left(12 \frac{\text{in}}{\text{ft}}\right)\left(3600 \frac{\text{sec}}{\text{hr}}\right)}$$
$$= 0.000395$$

The Reynolds number is less than 1, so Stokes' law applies.

If a particle of the maximum diameter starts at the water surface, the time required to settle out will be

$$t = \frac{d_s}{v_{s,\text{actual}}} = \frac{15 \text{ ft}}{0.71 \frac{\text{ft}}{\text{hr}}}$$
$$= 21.1 \text{ hr} \quad (20 \text{ hr})$$

The answer is (D).

231. The declining-growth method of population projection is

$$P = P_o + (P_s - P_o)(1 - e^{k\Delta t})$$

P_o is the base year population, P_s is the saturation population, and Δt is the future time period in years. The future population is

$$P = 100,000 + (300,000 - 100,000)(1 - e^{(-0.0055)(40)})$$
$$= 139,496$$

The current per capita demand is 150 gal/capita-day. The percent increase in population is

$$\left(\frac{139,496 - 100,000}{100,000}\right) \times 100\% = 39.5\%$$

The future per capita water demand will be

$$D_{\text{unit}} = (\text{per capita demand})(\text{decimal increase})$$
$$= \left(150 \frac{\text{gal}}{\text{capita-day}}\right)\left(\frac{100\% + (0.1)(39.5\%)}{100\%}\right)$$
$$= 155.9 \text{ gal/capita-day}$$

The future water demand will be

$$D_{\text{total}} = (\text{capita})(D_{\text{unit}})$$
$$= (139,496 \text{ capita})\left(155.9 \frac{\text{gal}}{\text{capita-day}}\right)$$
$$\times \left(\frac{1 \text{ MG}}{10^6 \text{ gal}}\right)$$
$$= 21.75 \text{ MGD} \quad (20 \text{ MGD})$$

The answer is (A).

232. Calculate the minimum required design capacity and compare it to the rated capacity for each unit process. The maximum daily design demand is

$$Q_{\text{max day}} = 1.5 Q_{\text{avg day}} = (1.5)\left(800 \ \frac{\text{L}}{\text{s}}\right)$$
$$= 1200 \ \text{L/s}$$

$$Q_{\text{max hour}} = 2.0 Q_{\text{max day}} = (2.0)\left(1200 \ \frac{\text{L}}{\text{s}}\right)$$
$$= 2400 \ \text{L/s}$$

The design capacities of the unit processes are given in the table.

unit process	design criteria	design criteria (L/s)
screening	1.25 maximum hourly demand	$1.25 \times 2400 = 3000$
influent flow meter	1.5 maximum hourly demand	$1.5 \times 2400 = 3600$
flash mix	1.25 maximum daily demand	$1.25 \times 1200 = 1500$
flocculation basin	1.25 maximum daily demand	$1.25 \times 1200 = 1500$
sedimentation basin	1.5 maximum daily demand	$1.5 \times 1200 = 1800$
filters	1.0 maximum daily demand	1200
effluent flow meter	1.5 maximum hourly demand	$1.5 \times 2400 = 3600$
chlorine contact/ clearwell	1.5 maximum daily demand	$1.5 \times 1200 = 1800$
high service pumps	1.0 maximum hourly demand	$1.0 \times 2400 = 2400$

The rated capacity of screening is

$$Q = \left(300\,000 \ \frac{\text{m}^3}{\text{d}}\right)\left(\frac{1 \ \text{d}}{86\,400 \ \text{s}}\right)\left(1000 \ \frac{\text{L}}{\text{m}^3}\right)$$
$$= 3472 \ \text{L/s}$$

The rated capacity of the influent flow meter is

$$Q = \left(14\,400 \ \frac{\text{m}^3}{\text{h}}\right)\left(\frac{1 \ \text{h}}{60 \ \text{min}}\right)\left(\frac{1 \ \text{min}}{60 \ \text{s}}\right)\left(1000 \ \frac{\text{L}}{\text{m}^3}\right)$$
$$= 4000 \ \text{L/s}$$

The rated capacity of flash mix is

$$Q = (20 \ \text{MGD})\left(10^6 \ \frac{\text{gal}}{\text{MG}}\right)\left(3.785 \ \frac{\text{L}}{\text{gal}}\right)\left(\frac{1 \ \text{d}}{86\,400 \ \text{s}}\right)$$
$$= 876 \ \text{L/s}$$

The rated capacity of the flocculation basin is

$$Q = \left(25{,}000 \ \frac{\text{gal}}{\text{min}}\right)\left(\frac{1 \ \text{min}}{60 \ \text{s}}\right)\left(3.785 \ \frac{\text{L}}{\text{gal}}\right)$$
$$= 1577 \ \text{L/s}$$

The rated capacity of the sedimentation basin is

$$Q = \left(2 \times 10^6 \ \frac{\text{gal}}{\text{h}}\right)\left(\frac{1 \ \text{h}}{60 \ \text{min}}\right)\left(\frac{1 \ \text{min}}{60 \ \text{s}}\right)\left(3.785 \ \frac{\text{L}}{\text{gal}}\right)$$
$$= 2103 \ \text{L/s}$$

The rated capacity of the filters is

$$Q = \left(50\,000 \ \frac{\text{L}}{\text{min}}\right)\left(\frac{1 \ \text{min}}{60 \ \text{s}}\right)$$
$$= 833 \ \text{L/s}$$

The rated capacity of the effluent flow meter is

$$Q = \left(10\,000 \ \frac{\text{m}^3}{\text{h}}\right)\left(\frac{1 \ \text{h}}{60 \ \text{min}}\right)\left(\frac{1 \ \text{min}}{60 \ \text{s}}\right)\left(\frac{1000 \ \text{L}}{\text{m}^3}\right)$$
$$= 2778 \ \text{L/s}$$

The rated capacity of the chlorine contact/clearwell is

$$Q = \left(70 \ \frac{\text{ft}^3}{\text{sec}}\right)\left(28.3 \ \frac{\text{L}}{\text{gal}}\right)$$
$$= 1981 \ \text{L/s}$$

The rated capacity of the high service pumps is

$$Q = \left(30{,}000 \ \frac{\text{gal}}{\text{min}}\right)\left(3.785 \ \frac{\text{L}}{\text{gal}}\right)\left(\frac{1 \ \text{min}}{60 \ \text{sec}}\right)$$
$$= 1892 \ \text{L/s}$$

A comparison of design criteria to rated capacities of the unit processes is given in the following table.

unit process	design criteria (L/s)	rated capacity (L/s)	OK?
screening	3000	3472	yes
influent flow meter	3600	4000	yes
flash mix	1500	876	no
flocculation basin	1500	1577	yes
sedimentation basin	1800	2103	yes
filters	1200	833	no
effluent flow meter	3600	2778	no
chlorine contact/clearwell	1800	1981	yes
high service pumps	2400	1892	no

The unit processes that require upgrading are flash mix, filters, effluent flow meter, and high service pumps.

The answer is (D).

233. For the first subduration, the demand consumption is

$$D = (\text{design demand rate})(\text{subduration})$$
$$= \left(995\,000 \ \frac{\text{m}^3}{\text{d}}\right)(7 \text{ d})$$
$$= 6\,965\,000 \text{ m}^3$$

The evaporation/seepage loss is

$$L = (\text{evaporation/seepage rate})(\text{reservoir area})$$
$$= (30 \text{ mm})\left(\frac{1 \text{ m}}{1000 \text{ mm}}\right)(4 \text{ km}^2)\left(1000 \ \frac{\text{m}}{\text{km}}\right)^2$$
$$= 120\,000 \text{ m}^3$$

The inflow volume is

$$I = (\text{average inflow rate})(\text{subduration})$$
$$= \left(120\,000 \ \frac{\text{m}^3}{\text{d}}\right)(7 \text{ d})$$
$$= 840\,000 \text{ m}^3$$

subduration (d)	average inflow rate (Mm^3/d)	inflow volume (Mm^3)	evaporation/ seepage rate (mm)
7	0.120	0.840	30
30	0.150	4.500	120
60	0.180	10.800	250
120	0.240	28.800	500
180	0.560	100.800	780
365	1.800	657.000	1600

subduration (d)	evaporation/ seepage volume (Mm^3)	design demand rate (Mm^3)	design demand volume (Mm^3)	required storage (Mm^3)
7	0.120	0.995	6.965	6.245
30	0.480	0.990	29.700	25.680
60	1.000	0.980	58.800	49.000
120	2.000	0.960	115.200	88.400
180	3.120	0.940	169.200	71.520
365	6.400	0.900	328.500	−322.100

The required storage volume can be determined from a mass balance with respect to the reservoir. A volume basis can be used because the density of water is essentially constant for these conditions. The required storage is the demand plus losses minus the inflow. The net storage required for the 7 d subduration is

$$S = D + L - I$$
$$= 6\,965\,000 \ \frac{\text{m}^3}{\text{d}} + 120\,000 \ \frac{\text{m}^3}{\text{d}} - 840\,000 \ \frac{\text{m}^3}{\text{d}}$$
$$= 6\,245\,000 \text{ m}^3$$

Calculations for the other subdurations are given in the table. The largest storage volume of the subdurations analyzed is $88\,400\,000 \text{ m}^3$ (88 Mm^3) for the 120 d subduration, which will contain the smaller volumes of other subdurations.

The answer is (D).

234. Statements I and IV are true.

Statement II is false. Hydrolytic reactions are not involved with cationic polymers, but such reactions are involved with anionic polymers.

Statement III is false. The pH is very important in most coagulant reactions.

Statement V is false. The detention time of in-line static mixers varies with the flow through the mixer.

The answer is (A).

235. The volume of water treated is

$$V = Qt$$
$$= \left(2 \ \frac{\text{m}^3}{\text{s}}\right)\left(86\,400 \ \frac{\text{s}}{\text{d}}\right)\left(1000 \ \frac{\text{L}}{\text{m}^3}\right)$$
$$= 172.8 \times 10^6 \text{ L/d}$$

The mass of alum sludge is

$$m_{\text{alum}} = QC(\text{efficiency})$$
$$= \left(172.8 \times 10^6 \ \frac{\text{L}}{\text{d}}\right)\left(12 \ \frac{\text{mg alum dose}}{\text{L}}\right)$$
$$\times \left(\frac{1 \text{ kg}}{10^6 \text{ mg}}\right)\left(0.46 \ \frac{\text{kg alum sludge}}{\text{kg alum dose}}\right)$$
$$\times (0.96)$$
$$= 915.7 \text{ kg/d alum sludge}$$

The equivalent concentration of TSS removed is

$$\Delta\text{TSS} = \Delta\text{NTU}^{1.2}$$
$$= (5 \text{ NTU} - 1 \text{ NTU})^{1.2}$$
$$= 5.28 \text{ mg/L}$$

The mass of TSS removed is

$$m_{\text{TSS}} = QC(\text{efficiency})$$

$$= \left(172.8 \times 10^6 \ \frac{\text{L}}{\text{d}}\right) \left(\frac{1 \ \text{kg}}{10^6 \ \text{mg}}\right) \left(5.28 \ \frac{\text{mg}}{\text{L}}\right)$$

$$\times (0.96)$$

$$= 875.9 \ \text{kg/d}$$

The mass of clay removed is

$$m_{\text{clay}} = QC(\text{efficiency})$$

$$= \left(172.8 \times 10^6 \ \frac{\text{L}}{\text{d}}\right) \left(\frac{1 \ \text{kg}}{10^6 \ \text{mg}}\right) \left(3 \ \frac{\text{mg}}{\text{L}}\right)$$

$$\times (0.96)$$

$$= 497.7 \ \text{kg/d}$$

The total mass of sludge generated following sedimentation is

$$m_{\text{sludge}} = m_{\text{alum}} + m_{\text{TSS}} + m_{\text{clay}}$$

$$= 915.7 \ \frac{\text{kg}}{\text{d}} + 875.9 \ \frac{\text{kg}}{\text{d}} + 497.7 \ \frac{\text{kg}}{\text{d}}$$

$$= 2289.3 \ \text{kg/d} \quad (2300 \ \text{kg/d})$$

The answer is (B).

236. Use the settling curves to determine the percent solids removal for the detention time associated with the midpoint between the curves. The 20% removal curve intersects the x-axis at 24 min.

The overflow rate is

$$V_o = \frac{\text{settling depth}}{\text{settling time}}$$

$$= \left(\frac{10 \ \text{ft}}{24 \ \text{min}}\right) \left(1440 \ \frac{\text{min}}{\text{day}}\right) \left(7.48 \ \frac{\text{gal}}{\text{ft}^3}\right)$$

$$= 4488 \ \text{gal/day-ft}^2$$

The detention time in hours is

$$t = (24 \ \text{min}) \left(\frac{1 \ \text{hr}}{60 \ \text{min}}\right)$$

$$= 0.40 \ \text{hr}$$

Project a line vertically from 24 min. The point midway between 20% and 30% removal (25%) corresponds to a depth of 6.4 ft. Similarly, the depths corresponding to other removal percentages for 0.4 hr detention are shown in the following table.

solids removal (%)	depth (ft)
25	6.4
35	3.8
45	2.8
55	2.4
65	2.0
75	1.7

The total percent removed for a detention time of 0.4 hr is

$$R_t = \text{percent removal for detention time}$$

$$+ \left(\frac{\text{depth at midpoint}}{\text{total depth}}\right)$$

$$\times (\text{increment of percent removal})$$

$$= 20\% + \left(\frac{6.4 \ \text{ft}}{10 \ \text{ft}}\right)(30\% - 20\%)$$

$$+ \left(\frac{3.8 \ \text{ft}}{10 \ \text{ft}}\right)(40\% - 30\%)$$

$$+ \left(\frac{2.8 \ \text{ft}}{10 \ \text{ft}}\right)(50\% - 40\%)$$

$$+ \left(\frac{2.4 \ \text{ft}}{10 \ \text{ft}}\right)(60\% - 50\%)$$

$$+ \left(\frac{2.0 \ \text{ft}}{10 \ \text{ft}}\right)(70\% - 60\%)$$

$$+ \left(\frac{1.7 \ \text{ft}}{10 \ \text{ft}}\right)(80\% - 70\%)$$

$$= 39.1\%$$

Calculations for the other removals and detention times are given in the following table.

detention time (hr)	overflow rate (gal/day-ft^2)	solids removal (%)
0.40	4488	39.1
0.63	2835	52.5
0.87	2071	63.7
1.07	1683	70.4
1.30	1381	75.6
1.57	1146	79.1

For 75% removal, the required detention time is 1.3 hr and the overflow rate (OFR) is 1380 gal/day-ft^2 (scale-up factor is 1.0).

The required area is

$$A = \frac{Q}{\text{OFR}}$$

$$= \frac{1.5 \times 10^6 \ \frac{\text{gal}}{\text{day}}}{1380 \ \frac{\text{gal}}{\text{day-ft}^2}}$$

$$= 1087 \ \text{ft}^2$$

The diameter is

$$D = \sqrt{\frac{4A}{\pi}}$$

$$= \sqrt{\frac{(4)(1087 \ \text{ft}^2)}{\pi}}$$

$$= 37.2 \ \text{ft} \quad (37 \ \text{ft})$$

The required depth is

$$h = \frac{Qt}{A}$$

$$= \frac{\left(1.5 \times 10^6 \frac{\text{gal}}{\text{day}}\right)(1.3 \text{ hr})\left(\frac{1 \text{ day}}{24 \text{ hr}}\right)\left(\frac{1 \text{ ft}^3}{7.48 \text{ gal}}\right)}{1087 \text{ ft}^2}$$

$$= 9.99 \text{ ft} \quad (10 \text{ ft})$$

The answer is (B).

237. Since each layer is uniform, the weight fraction of each layer is 1, and all particles in the layer have the same diameter. The head loss can be found from the Rose equation. Sum the head loss from the individual layers to find the total head loss. The kinematic viscosity at 10°C is $1.371 \times 10^{-6} \text{ m}^2/\text{s}$. The approach velocity is the filtration rate as velocity.

$$v_a = \left(4.5 \frac{\text{L}}{\text{s·m}^2}\right)\left(\frac{1 \text{ m}^3}{1000 \text{ L}}\right)$$

$$= 0.0045 \text{ m/s}$$

For the sand layer, the Reynolds number is

$$\text{Re} = \frac{\phi d v_a}{\nu}$$

$$= \frac{(0.8)(1.0 \text{ mm})\left(\frac{1 \text{ m}}{1000 \text{ mm}}\right)\left(0.0045 \frac{\text{m}}{\text{s}}\right)}{1.371 \times 10^{-6} \frac{\text{m}^2}{\text{s}}}$$

$$= 2.626$$

Using the equation developed by T.R. Camp to describe drag on spheres. For a Reynolds number greater than 1 but less than 10,000, the drag coefficient is

$$C_D = \frac{24}{\text{Re}} + \frac{3}{\sqrt{\text{Re}}} + 0.34$$

$$= \frac{24}{2.626} + \frac{3}{\sqrt{2.626}} + 0.34$$

$$= 11.33$$

Using the Rose equation for stratified beds with uniform porosity, the head loss in the sand layer is

$$h_L = \left(\frac{1.067 D v_a^2}{\phi g \xi^4}\right)\sum \frac{C_D x}{d}$$

D is the depth of weight fraction, and x is the weight fraction of particles (1 for uniform layer).

$$h_L = \frac{(1.067)(280 \text{ mm})\left(\frac{1 \text{ m}}{1000 \text{ mm}}\right) \times \left(0.0045 \frac{\text{m}}{\text{s}}\right)^2 (11.33)(1)}{(0.8)\left(9.81 \frac{\text{m}}{\text{s}^2}\right)(0.50)^4(1.0 \text{ mm})\left(\frac{1 \text{ m}}{1000 \text{ mm}}\right)}$$

$$= 0.1397 \text{ m} \quad (0.140 \text{ m})$$

For the garnet layer, the Reynolds number is

$$\text{Re} = \frac{(0.8)(0.40 \text{ mm})\left(\frac{1 \text{ m}}{1000 \text{ mm}}\right)\left(0.0045 \frac{\text{m}}{\text{s}}\right)}{1.371 \times 10^{-6} \frac{\text{m}^2}{\text{s}}}$$

$$= 1.05$$

For a Reynolds number greater than 1 but less than 10,000, the drag coefficient is

$$C_D = \frac{24}{1.05} + \frac{3}{\sqrt{1.05}} + 0.34$$

$$= 26.12$$

The head loss in the garnet layer is

$$h_L = \frac{(1.067)(100 \text{ mm})\left(\frac{1 \text{ m}}{1000 \text{ mm}}\right) \times \left(0.0045 \frac{\text{m}}{\text{s}}\right)^2 (26.12)(1)}{(0.8)\left(9.81 \frac{\text{m}}{\text{s}^2}\right)(0.55)^4(0.40 \text{ mm})\left(\frac{1 \text{ m}}{1000 \text{ mm}}\right)}$$

$$= 0.1965 \text{ m} \quad (0.20 \text{ m})$$

The total head loss is

$$\sum h_L = 0.14 \text{ m} + 0.20 \text{ m} = 0.34 \text{ m}$$

The answer is (D).

238. The total volume of water to be disinfected is

$$V = (1000 \text{ m}^3)\left(1000 \frac{\text{L}}{\text{m}^3}\right)\left(\frac{1 \text{ ML}}{10^6 \text{ L}}\right)$$

$$= 1 \text{ ML}$$

The mass of water to be treated is

$$m_{\text{water}} = (1 \text{ ML})\left(10^6 \frac{\text{L}}{\text{ML}}\right)\left(1 \frac{\text{kg}}{\text{L}}\right)$$

$$\times \left(1000 \frac{\text{g}}{\text{kg}}\right)\left(\frac{1 \text{ Gg}}{10^9 \text{ g}}\right)$$

$$= 1 \text{ Gg}$$

The mass of hypochlorite required is

$$m_{\text{hypo}} = m_{\text{water}} C_{\text{hypo}}$$

$$= (1 \text{ Gg})\left(\frac{0.5 \text{ kg OCl}}{100 \text{ kg solution}}\right)\left(\frac{1 \text{ kg solution}}{0.7 \text{ kg OCl}}\right)$$

$$\times \left(\frac{1 \text{ Gg}}{10^9 \text{ g}}\right)\left(10^9 \frac{\text{g}}{\text{Gg}}\right)\left(\frac{1 \text{ kg}}{1000 \text{ g}}\right)$$

$$= 7143 \text{ kg} \quad (7000 \text{ kg})$$

The answer is (D).

239. Statements I, III, and V are true.

Statement II is false. Enforcement of primary standards by states is mandatory.

Statement IV is false. The MCL for turbidity is 0.5 NTU in a maximum of 5% of monthly samples.

Statement VI is false. Microbiological contamination shall not exceed 5% positive samples.

The answer is (C).

240. The log-log relationship is

$$\ln F = m \ln t + k$$

F is the flux in gal/day-ft^2. m is the slope of the line, and k is the constant. Solve for the slope.

$$m = \frac{\ln 24.2 - \ln 7.0}{\ln 10 - \ln 10{,}000}$$
$$= -0.18$$

Check for the next increment.

$$m = \frac{\ln 7.0 - \ln 6.2}{\ln 10{,}000 - \ln 20{,}000}$$
$$= -0.18$$

Solve for the constant.

$$k = \ln 7.0 + 0.18 \ln 10{,}000$$
$$= 3.60$$

The equation is

$$\ln F = -0.18 \ln t + 3.60$$

At 2 yr, flux is

$$\ln F = -0.18 \ln \left((2 \text{ yr}) \left(365 \ \frac{\text{day}}{\text{yr}} \right) \left(24 \ \frac{\text{hr}}{\text{day}} \right) \right)$$
$$+ 3.60$$
$$= 1.841$$
$$F = e^{1.841} = 6.30 \text{ gal/ft}^2\text{-day}$$

At 3 yr, flux is

$$\ln F = -0.18 \ln \left((3 \text{ yr}) \left(365 \ \frac{\text{day}}{\text{yr}} \right) \left(24 \ \frac{\text{hr}}{\text{day}} \right) \right)$$
$$+ 3.60$$
$$= 1.768$$
$$F = e^{1.768} = 5.86 \text{ gal/ft}^2\text{-day}$$

The percent reduction is

$$\% \text{ reduction} = \left(\frac{6.30 \ \frac{\text{gal}}{\text{ft}^2\text{-day}} - 5.86 \ \frac{\text{gal}}{\text{ft}^2\text{-day}}}{6.30 \ \frac{\text{gal}}{\text{ft}^2\text{-day}}} \right)$$
$$\times 100\%$$
$$= 6.98\% \quad (7\%)$$

The answer is (B).

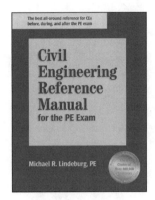
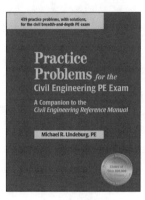

Email Updates Keep You on Top of Your Exam

You need current information to be fully prepared for your exam. Register for PPI's Email Updates to receive convenient updates relevant to the specific exam you are taking. Our updates include notices of exam changes, useful exam tips, errata postings, and new product announcements. There is no charge for this service, and you can cancel at any time.

Register at **www.ppi2pass.com/cgi-bin/signup.cgi**

Free Catalog of Tried-and-True Exam Products

Get a free PPI catalog with a comprehensive selection of the best FE, PE, SE, FLS, and PLS exam-review products available, user tested by more than 800,000 engineers and surveyors. Included are books, software, videos, and the NCEES sample-question books.

Request a catalog at **www.ppi2pass.com/catalogrequest**

How to Report Errors

Find an error? You can report it in two easy steps.

First, check the errata listings on our website, at **www.ppi2pass.com/errata**. The item you noticed may already have been identified. It's always a good idea to check this page before you start studying and periodically thereafter.

Then, go to PPI's Errata Report Form at **www.ppi2pass.com/erratasubmit**, and tell us about the discrepancy you think you've found. Your information will be forwarded to the appropriate author or subject matter expert for verification. Valid corrections are added to the errata section of our website.

You may also fax errata to us at 650-592-4519 or mail them to Professional Publications, Inc., c/o Editorial Errata Department, 1250 Fifth Ave., Belmont, CA 94002. Be sure to include your name, the book title, the edition and printing numbers, the page number(s), and any other information that will help us locate the error(s).

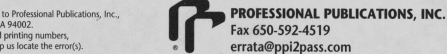

PROFESSIONAL PUBLICATIONS, INC.
Fax 650-592-4519
errata@ppi2pass.com